Rocks: A Very Short Introduction

VERY SHORT INTRODUCTIONS are for anyone wanting a stimulating and accessible way into a new subject. They are written by experts, and have been translated into more than 40 different languages.

The series began in 1995, and now covers a wide variety of topics in every discipline. The VSI library now contains over 500 volumes—a Very Short Introduction to everything from Psychology and Philosophy of Science to American History and Relativity—and continues to grow in every subject area.

Very Short Introductions available now:

ACCOUNTING Christopher Nobes
ADOLESCENCE Peter K. Smith
ADVERTISING Winston Fletcher
AFRICAN AMERICAN RELIGION Eddie S. Glaude Jr
AFRICAN HISTORY John Parker and Richard Rathbone
AFRICAN RELIGIONS Jacob K. Olupona
AGEING Nancy A. Pachana
AGNOSTICISM Robin Le Poidevin
AGRICULTURE Paul Brassley and Richard Soffe
ALEXANDER THE GREAT Hugh Bowden
ALGEBRA Peter M. Higgins
AMERICAN HISTORY Paul S. Boyer
AMERICAN IMMIGRATION David A. Gerber
AMERICAN LEGAL HISTORY G. Edward White
AMERICAN POLITICAL HISTORY Donald Critchlow
AMERICAN POLITICAL PARTIES AND ELECTIONS L. Sandy Maisel
AMERICAN POLITICS Richard M. Valelly
THE AMERICAN PRESIDENCY Charles O. Jones
THE AMERICAN REVOLUTION Robert J. Allison
AMERICAN SLAVERY Heather Andrea Williams
THE AMERICAN WEST Stephen Aron
AMERICAN WOMEN'S HISTORY Susan Ware
ANAESTHESIA Aidan O'Donnell
ANARCHISM Colin Ward
ANCIENT ASSYRIA Karen Radner
ANCIENT EGYPT Ian Shaw
ANCIENT EGYPTIAN ART AND ARCHITECTURE Christina Riggs
ANCIENT GREECE Paul Cartledge
THE ANCIENT NEAR EAST Amanda H. Podany
ANCIENT PHILOSOPHY Julia Annas
ANCIENT WARFARE Harry Sidebottom
ANGELS David Albert Jones
ANGLICANISM Mark Chapman
THE ANGLO-SAXON AGE John Blair
THE ANIMAL KINGDOM Peter Holland
ANIMAL RIGHTS David DeGrazia
THE ANTARCTIC Klaus Dodds
ANTISEMITISM Steven Beller
ANXIETY Daniel Freeman and Jason Freeman
THE APOCRYPHAL GOSPELS Paul Foster
ARCHAEOLOGY Paul Bahn
ARCHITECTURE Andrew Ballantyne
ARISTOCRACY William Doyle
ARISTOTLE Jonathan Barnes
ART HISTORY Dana Arnold
ART THEORY Cynthia Freeland
ASIAN AMERICAN HISTORY Madeline Y. Hsu
ASTROBIOLOGY David C. Catling
ASTROPHYSICS James Binney
ATHEISM Julian Baggini
AUGUSTINE Henry Chadwick

AUSTRALIA Kenneth Morgan
AUTISM Uta Frith
THE AVANT GARDE David Cottington
THE AZTECS Davíd Carrasco
BABYLONIA Trevor Bryce
BACTERIA Sebastian G. B. Amyes
BARTHES Jonathan Culler
THE BEATS David Sterritt
BEAUTY Roger Scruton
BESTSELLERS John Sutherland
THE BIBLE John Riches
BIBLICAL ARCHAEOLOGY Eric H. Cline
BIOGRAPHY Hermione Lee
BLACK HOLES Katherine Blundell
BLOOD Chris Cooper
THE BLUES Elijah Wald
THE BODY Chris Shilling
THE BOOK OF MORMON Terryl Givens
BORDERS Alexander C. Diener and Joshua Hagen
THE BRAIN Michael O'Shea
THE BRICS Andrew F. Cooper
THE BRITISH CONSTITUTION Martin Loughlin
THE BRITISH EMPIRE Ashley Jackson
BRITISH POLITICS Anthony Wright
BUDDHA Michael Carrithers
BUDDHISM Damien Keown
BUDDHIST ETHICS Damien Keown
BYZANTIUM Peter Sarris
CALVINISM Jon Balserak
CANCER Nicholas James
CAPITALISM James Fulcher
CATHOLICISM Gerald O'Collins
CAUSATION Stephen Mumford and Rani Lill Anjum
THE CELL Terence Allen and Graham Cowling
THE CELTS Barry Cunliffe
CHAOS Leonard Smith
CHEMISTRY Peter Atkins
CHILD PSYCHOLOGY Usha Goswami
CHILDREN'S LITERATURE Kimberley Reynolds
CHINESE LITERATURE Sabina Knight
CHOICE THEORY Michael Allingham
CHRISTIAN ART Beth Williamson
CHRISTIAN ETHICS D. Stephen Long
CHRISTIANITY Linda Woodhead
CITIZENSHIP Richard Bellamy
CIVIL ENGINEERING David Muir Wood
CLASSICAL LITERATURE William Allan
CLASSICAL MYTHOLOGY Helen Morales
CLASSICS Mary Beard and John Henderson
CLAUSEWITZ Michael Howard
CLIMATE Mark Maslin
CLIMATE CHANGE Mark Maslin
COGNITIVE NEUROSCIENCE Richard Passingham
THE COLD WAR Robert McMahon
COLONIAL AMERICA Alan Taylor
COLONIAL LATIN AMERICAN LITERATURE Rolena Adorno
COMBINATORICS Robin Wilson
COMEDY Matthew Bevis
COMMUNISM Leslie Holmes
COMPLEXITY John H. Holland
THE COMPUTER Darrel Ince
COMPUTER SCIENCE Subrata Dasgupta
CONFUCIANISM Daniel K. Gardner
THE CONQUISTADORS Matthew Restall and Felipe Fernández-Armesto
CONSCIENCE Paul Strohm
CONSCIOUSNESS Susan Blackmore
CONTEMPORARY ART Julian Stallabrass
CONTEMPORARY FICTION Robert Eaglestone
CONTINENTAL PHILOSOPHY Simon Critchley
COPERNICUS Owen Gingerich
CORAL REEFS Charles Sheppard
CORPORATE SOCIAL RESPONSIBILITY Jeremy Moon
CORRUPTION Leslie Holmes
COSMOLOGY Peter Coles
CRIME FICTION Richard Bradford
CRIMINAL JUSTICE Julian V. Roberts
CRITICAL THEORY Stephen Eric Bronner
THE CRUSADES Christopher Tyerman
CRYPTOGRAPHY Fred Piper and Sean Murphy
CRYSTALLOGRAPHY A. M. Glazer
THE CULTURAL REVOLUTION Richard Curt Kraus

DADA AND SURREALISM David Hopkins
DANTE Peter Hainsworth and David Robey
DARWIN Jonathan Howard
THE DEAD SEA SCROLLS Timothy Lim
DECOLONIZATION Dane Kennedy
DEMOCRACY Bernard Crick
DERRIDA Simon Glendinning
DESCARTES Tom Sorell
DESERTS Nick Middleton
DESIGN John Heskett
DEVELOPMENTAL BIOLOGY Lewis Wolpert
THE DEVIL Darren Oldridge
DIASPORA Kevin Kenny
DICTIONARIES Lynda Mugglestone
DINOSAURS David Norman
DIPLOMACY Joseph M. Siracusa
DOCUMENTARY FILM Patricia Aufderheide
DREAMING J. Allan Hobson
DRUGS Les Iversen
DRUIDS Barry Cunliffe
EARLY MUSIC Thomas Forrest Kelly
THE EARTH Martin Redfern
EARTH SYSTEM SCIENCE Tim Lenton
ECONOMICS Partha Dasgupta
EDUCATION Gary Thomas
EGYPTIAN MYTH Geraldine Pinch
EIGHTEENTH-CENTURY BRITAIN Paul Langford
THE ELEMENTS Philip Ball
EMOTION Dylan Evans
EMPIRE Stephen Howe
ENGELS Terrell Carver
ENGINEERING David Blockley
ENGLISH LITERATURE Jonathan Bate
THE ENLIGHTENMENT John Robertson
ENTREPRENEURSHIP Paul Westhead and Mike Wright
ENVIRONMENTAL ECONOMICS Stephen Smith
ENVIRONMENTAL POLITICS Andrew Dobson
EPICUREANISM Catherine Wilson
EPIDEMIOLOGY Rodolfo Saracci
ETHICS Simon Blackburn
ETHNOMUSICOLOGY Timothy Rice
THE ETRUSCANS Christopher Smith
EUGENICS Philippa Levine
THE EUROPEAN UNION John Pinder and Simon Usherwood
EVOLUTION Brian and Deborah Charlesworth
EXISTENTIALISM Thomas Flynn
EXPLORATION Stewart A. Weaver
THE EYE Michael Land
FAMILY LAW Jonathan Herring
FASCISM Kevin Passmore
FASHION Rebecca Arnold
FEMINISM Margaret Walters
FILM Michael Wood
FILM MUSIC Kathryn Kalinak
THE FIRST WORLD WAR Michael Howard
FOLK MUSIC Mark Slobin
FOOD John Krebs
FORENSIC PSYCHOLOGY David Canter
FORENSIC SCIENCE Jim Fraser
FORESTS Jaboury Ghazoul
FOSSILS Keith Thomson
FOUCAULT Gary Gutting
THE FOUNDING FATHERS R. B. Bernstein
FRACTALS Kenneth Falconer
FREE SPEECH Nigel Warburton
FREE WILL Thomas Pink
FRENCH LITERATURE John D. Lyons
THE FRENCH REVOLUTION William Doyle
FREUD Anthony Storr
FUNDAMENTALISM Malise Ruthven
FUNGI Nicholas P. Money
GALAXIES John Gribbin
GALILEO Stillman Drake
GAME THEORY Ken Binmore
GANDHI Bhikhu Parekh
GENES Jonathan Slack
GENIUS Andrew Robinson
GEOGRAPHY John Matthews and David Herbert
GEOPOLITICS Klaus Dodds
GERMAN LITERATURE Nicholas Boyle
GERMAN PHILOSOPHY Andrew Bowie
GLOBAL CATASTROPHES Bill McGuire

GLOBAL ECONOMIC HISTORY Robert C. Allen
GLOBALIZATION Manfred Steger
GOD John Bowker
GOETHE Ritchie Robertson
THE GOTHIC Nick Groom
GOVERNANCE Mark Bevir
THE GREAT DEPRESSION AND THE NEW DEAL Eric Rauchway
HABERMAS James Gordon Finlayson
HAPPINESS Daniel M. Haybron
THE HARLEM RENAISSANCE Cheryl A. Wall
THE HEBREW BIBLE AS LITERATURE Tod Linafelt
HEGEL Peter Singer
HEIDEGGER Michael Inwood
HERMENEUTICS Jens Zimmermann
HERODOTUS Jennifer T. Roberts
HIEROGLYPHS Penelope Wilson
HINDUISM Kim Knott
HISTORY John H. Arnold
THE HISTORY OF ASTRONOMY Michael Hoskin
THE HISTORY OF CHEMISTRY William H. Brock
THE HISTORY OF LIFE Michael Benton
THE HISTORY OF MATHEMATICS Jacqueline Stedall
THE HISTORY OF MEDICINE William Bynum
THE HISTORY OF TIME Leofranc Holford-Strevens
HIV AND AIDS Alan Whiteside
HOBBES Richard Tuck
HOLLYWOOD Peter Decherney
HOME Michael Allen Fox
HORMONES Martin Luck
HUMAN ANATOMY Leslie Klenerman
HUMAN EVOLUTION Bernard Wood
HUMAN RIGHTS Andrew Clapham
HUMANISM Stephen Law
HUME A. J. Ayer
HUMOUR Noël Carroll
THE ICE AGE Jamie Woodward
IDEOLOGY Michael Freeden
INDIAN CINEMA Ashish Rajadhyaksha
INDIAN PHILOSOPHY Sue Hamilton
INFECTIOUS DISEASE Marta L. Wayne and Benjamin M. Bolker
INFORMATION Luciano Floridi
INNOVATION Mark Dodgson and David Gann
INTELLIGENCE Ian J. Deary
INTERNATIONAL LAW Vaughan Lowe
INTERNATIONAL MIGRATION Khalid Koser
INTERNATIONAL RELATIONS Paul Wilkinson
INTERNATIONAL SECURITY Christopher S. Browning
IRAN Ali M. Ansari
ISLAM Malise Ruthven
ISLAMIC HISTORY Adam Silverstein
ISOTOPES Rob Ellam
ITALIAN LITERATURE Peter Hainsworth and David Robey
JESUS Richard Bauckham
JOURNALISM Ian Hargreaves
JUDAISM Norman Solomon
JUNG Anthony Stevens
KABBALAH Joseph Dan
KAFKA Ritchie Robertson
KANT Roger Scruton
KEYNES Robert Skidelsky
KIERKEGAARD Patrick Gardiner
KNOWLEDGE Jennifer Nagel
THE KORAN Michael Cook
LANDSCAPE ARCHITECTURE Ian H. Thompson
LANDSCAPES AND GEOMORPHOLOGY Andrew Goudie and Heather Viles
LANGUAGES Stephen R. Anderson
LATE ANTIQUITY Gillian Clark
LAW Raymond Wacks
THE LAWS OF THERMODYNAMICS Peter Atkins
LEADERSHIP Keith Grint
LEARNING Mark Haselgrove
LEIBNIZ Maria Rosa Antognazza
LIBERALISM Michael Freeden
LIGHT Ian Walmsley
LINCOLN Allen C. Guelzo
LINGUISTICS Peter Matthews
LITERARY THEORY Jonathan Culler
LOCKE John Dunn
LOGIC Graham Priest

LOVE Ronald de Sousa
MACHIAVELLI Quentin Skinner
MADNESS Andrew Scull
MAGIC Owen Davies
MAGNA CARTA Nicholas Vincent
MAGNETISM Stephen Blundell
MALTHUS Donald Winch
MANAGEMENT John Hendry
MAO Delia Davin
MARINE BIOLOGY Philip V. Mladenov
THE MARQUIS DE SADE John Phillips
MARTIN LUTHER Scott H. Hendrix
MARTYRDOM Jolyon Mitchell
MARX Peter Singer
MATERIALS Christopher Hall
MATHEMATICS Timothy Gowers
THE MEANING OF LIFE
Terry Eagleton
MEASUREMENT David Hand
MEDICAL ETHICS Tony Hope
MEDICAL LAW Charles Foster
MEDIEVAL BRITAIN John Gillingham
and Ralph A. Griffiths
MEDIEVAL LITERATURE
Elaine Treharne
MEDIEVAL PHILOSOPHY
John Marenbon
MEMORY Jonathan K. Foster
METAPHYSICS Stephen Mumford
THE MEXICAN REVOLUTION
Alan Knight
MICHAEL FARADAY
Frank A. J. L. James
MICROBIOLOGY Nicholas P. Money
MICROECONOMICS Avinash Dixit
MICROSCOPY Terence Allen
THE MIDDLE AGES Miri Rubin
MILITARY JUSTICE Eugene R. Fidell
MINERALS David Vaughan
MODERN ART David Cottington
MODERN CHINA Rana Mitter
MODERN DRAMA
Kirsten E. Shepherd-Barr
MODERN FRANCE
Vanessa R. Schwartz
MODERN IRELAND Senia Pašeta
MODERN ITALY Anna Cento Bull
MODERN JAPAN
Christopher Goto-Jones
MODERN LATIN AMERICAN
LITERATURE
Roberto González Echevarría
MODERN WAR Richard English
MODERNISM Christopher Butler
MOLECULAR BIOLOGY
Aysha Divan and Janice A. Royds
MOLECULES Philip Ball
THE MONGOLS Morris Rossabi
MOONS David A. Rothery
MORMONISM
Richard Lyman Bushman
MOUNTAINS Martin F. Price
MUHAMMAD Jonathan A. C. Brown
MULTICULTURALISM
Ali Rattansi
MUSIC Nicholas Cook
MYTH Robert A. Segal
THE NAPOLEONIC WARS
Mike Rapport
NATIONALISM Steven Grosby
NELSON MANDELA Elleke Boehmer
NEOLIBERALISM Manfred Steger and
Ravi Roy
NETWORKS Guido Caldarelli and
Michele Catanzaro
THE NEW TESTAMENT
Luke Timothy Johnson
THE NEW TESTAMENT AS
LITERATURE Kyle Keefer
NEWTON Robert Iliffe
NIETZSCHE Michael Tanner
NINETEENTH-CENTURY BRITAIN
Christopher Harvie and
H. C. G. Matthew
THE NORMAN CONQUEST
George Garnett
NORTH AMERICAN INDIANS
Theda Perdue and Michael D. Green
NORTHERN IRELAND
Marc Mulholland
NOTHING Frank Close
NUCLEAR PHYSICS Frank Close
NUCLEAR POWER Maxwell Irvine
NUCLEAR WEAPONS
Joseph M. Siracusa
NUMBERS Peter M. Higgins
NUTRITION David A. Bender
OBJECTIVITY Stephen Gaukroger
THE OLD TESTAMENT
Michael D. Coogan
THE ORCHESTRA D. Kern Holoman
ORGANIZATIONS Mary Jo Hatch
PANDEMICS Christian W. McMillen
PAGANISM Owen Davies

THE PALESTINIAN-ISRAELI CONFLICT Martin Bunton
PARTICLE PHYSICS Frank Close
PAUL E. P. Sanders
PEACE Oliver P. Richmond
PENTECOSTALISM William K. Kay
THE PERIODIC TABLE Eric R. Scerri
PHILOSOPHY Edward Craig
PHILOSOPHY IN THE ISLAMIC WORLD Peter Adamson
PHILOSOPHY OF LAW Raymond Wacks
PHILOSOPHY OF SCIENCE Samir Okasha
PHOTOGRAPHY Steve Edwards
PHYSICAL CHEMISTRY Peter Atkins
PILGRIMAGE Ian Reader
PLAGUE Paul Slack
PLANETS David A. Rothery
PLANTS Timothy Walker
PLATE TECTONICS Peter Molnar
PLATO Julia Annas
POLITICAL PHILOSOPHY David Miller
POLITICS Kenneth Minogue
POSTCOLONIALISM Robert Young
POSTMODERNISM Christopher Butler
POSTSTRUCTURALISM Catherine Belsey
PREHISTORY Chris Gosden
PRESOCRATIC PHILOSOPHY Catherine Osborne
PRIVACY Raymond Wacks
PROBABILITY John Haigh
PROGRESSIVISM Walter Nugent
PROTESTANTISM Mark A. Noll
PSYCHIATRY Tom Burns
PSYCHOANALYSIS Daniel Pick
PSYCHOLOGY Gillian Butler and Freda McManus
PSYCHOTHERAPY Tom Burns and Eva Burns-Lundgren
PUBLIC ADMINISTRATION Stella Z. Theodoulou and Ravi K. Roy
PUBLIC HEALTH Virginia Berridge
PURITANISM Francis J. Bremer
THE QUAKERS Pink Dandelion
QUANTUM THEORY John Polkinghorne
RACISM Ali Rattansi
RADIOACTIVITY Claudio Tuniz
RASTAFARI Ennis B. Edmonds
THE REAGAN REVOLUTION Gil Troy
REALITY Jan Westerhoff
THE REFORMATION Peter Marshall
RELATIVITY Russell Stannard
RELIGION IN AMERICA Timothy Beal
THE RENAISSANCE Jerry Brotton
RENAISSANCE ART Geraldine A. Johnson
REVOLUTIONS Jack A. Goldstone
RHETORIC Richard Toye
RISK Baruch Fischhoff and John Kadvany
RITUAL Barry Stephenson
RIVERS Nick Middleton
ROBOTICS Alan Winfield
ROCKS Jan Zalasiewicz
ROMAN BRITAIN Peter Salway
THE ROMAN EMPIRE Christopher Kelly
THE ROMAN REPUBLIC David M. Gwynn
ROMANTICISM Michael Ferber
ROUSSEAU Robert Wokler
RUSSELL A. C. Grayling
RUSSIAN HISTORY Geoffrey Hosking
RUSSIAN LITERATURE Catriona Kelly
THE RUSSIAN REVOLUTION S. A. Smith
SAVANNAS Peter A. Furley
SCHIZOPHRENIA Chris Frith and Eve Johnstone
SCHOPENHAUER Christopher Janaway
SCIENCE AND RELIGION Thomas Dixon
SCIENCE FICTION David Seed
THE SCIENTIFIC REVOLUTION Lawrence M. Principe
SCOTLAND Rab Houston
SEXUALITY Véronique Mottier
SHAKESPEARE'S COMEDIES Bart van Es
SIKHISM Eleanor Nesbitt
THE SILK ROAD James A. Millward
SLANG Jonathon Green
SLEEP Steven W. Lockley and Russell G. Foster
SOCIAL AND CULTURAL ANTHROPOLOGY John Monaghan and Peter Just
SOCIAL PSYCHOLOGY Richard J. Crisp

SOCIAL WORK Sally Holland and Jonathan Scourfield
SOCIALISM Michael Newman
SOCIOLINGUISTICS John Edwards
SOCIOLOGY Steve Bruce
SOCRATES C. C. W. Taylor
SOUND Mike Goldsmith
THE SOVIET UNION Stephen Lovell
THE SPANISH CIVIL WAR Helen Graham
SPANISH LITERATURE Jo Labanyi
SPINOZA Roger Scruton
SPIRITUALITY Philip Sheldrake
SPORT Mike Cronin
STARS Andrew King
STATISTICS David J. Hand
STEM CELLS Jonathan Slack
STRUCTURAL ENGINEERING David Blockley
STUART BRITAIN John Morrill
SUPERCONDUCTIVITY Stephen Blundell
SYMMETRY Ian Stewart
TAXATION Stephen Smith
TEETH Peter S. Ungar
TELESCOPES Geoff Cottrell
TERRORISM Charles Townshend
THEATRE Marvin Carlson
THEOLOGY David F. Ford
THOMAS AQUINAS Fergus Kerr
THOUGHT Tim Bayne
TIBETAN BUDDHISM Matthew T. Kapstein
TOCQUEVILLE Harvey C. Mansfield
TRAGEDY Adrian Poole
TRANSLATION Matthew Reynolds
THE TROJAN WAR Eric H. Cline
TRUST Katherine Hawley
THE TUDORS John Guy
TWENTIETH-CENTURY BRITAIN Kenneth O. Morgan
THE UNITED NATIONS Jussi M. Hanhimäki
THE U.S. CONGRESS Donald A. Ritchie
THE U.S. SUPREME COURT Linda Greenhouse
UTOPIANISM Lyman Tower Sargent
THE VIKINGS Julian Richards
VIRUSES Dorothy H. Crawford
WAR AND TECHNOLOGY Alex Roland
WATER John Finney
THE WELFARE STATE David Garland
WILLIAM SHAKESPEARE Stanley Wells
WITCHCRAFT Malcolm Gaskill
WITTGENSTEIN A. C. Grayling
WORK Stephen Fineman
WORLD MUSIC Philip Bohlman
THE WORLD TRADE ORGANIZATION Amrita Narlikar
WORLD WAR II Gerhard L. Weinberg
WRITING AND SCRIPT Andrew Robinson
ZIONISM Michael Stanislawski

Available soon:

BANKING John Goddard and John O. S. Wilson
ASIAN AMERICAN HISTORY Madeline Y. Hsu
PANDEMICS Christian W. McMillen
ZIONISM Michael Stanislawski
THE INDUSTRIAL REVOLUTION Robert C. Allen

Jan Zalasiewicz

ROCKS

A Very Short Introduction

OXFORD
UNIVERSITY PRESS

OXFORD
UNIVERSITY PRESS

Great Clarendon Street, Oxford, OX2 6DP,
United Kingdom

Oxford University Press is a department of the University of Oxford.
It furthers the University's objective of excellence in research, scholarship,
and education by publishing worldwide. Oxford is a registered trade mark of
Oxford University Press in the UK and in certain other countries

First edition published in 2016

Published in the United States of America by Oxford University Press
198 Madison Avenue, New York, NY 10016, United States of America

British Library Cataloguing in Publication Data
Data available

Library of Congress Control Number: 2016951327

ISBN 978–0–19–872519–0

Printed and bound by
CPI Group (UK) Ltd, Croydon, CR0 4YY

Dedicated to the memory of Ryszard Kryza of the University of Wrocław, Poland

Contents

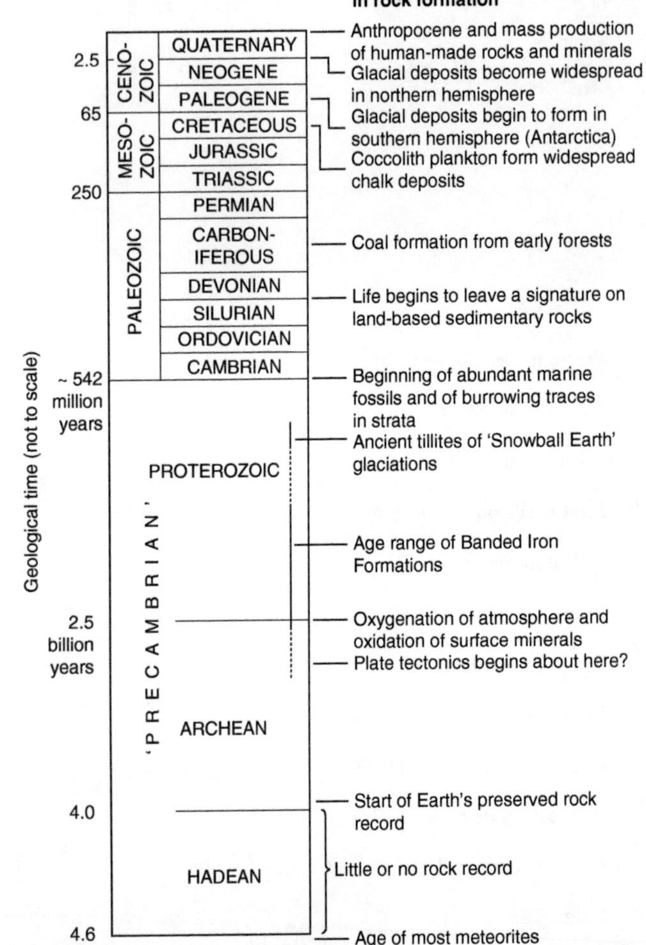

1. GEOLOGICAL TIME SCALE.

Prologue

Rocks, more than anything else, underpin our lives. They make up the solid structure of the Earth and of other rocky planets, and are present at the cores of gas giant planets. We live on the rocky surface of the planet, grow our food on weathered debris derived from rocks, and from rocks we obtain nearly all of the raw materials with which we build our civilization. Increasingly, we are making rocks on a planetary scale, in the form of gigantic amounts of concrete, brick, and ceramic. More widely, rocks contain our sense of planetary history: indeed, in a very literal sense they *are* the evidence from which Earth history, as encapsulated in the GEOLOGICAL TIME SCALE (Figure 1), is constructed. And, in their guise as petrified history, rocks are a guide to our future, too.

This VSI will give some flavour of the scale, structure, and diversity of rocks on Earth as well as in outer space and on other planets. It will consider how rocks act as planetary foundations—even planetary regulators. It will consider how rocks are formed, how they evolve within enormous cycles of transformation, and how we examine them and glean histories from them. There is almost an infinite variety of rocks and, properly interrogated, they can tell an almost infinite number of stories. These pages, hopefully, will represent some sort of starting point.

My grateful thanks to: Latha Menon and the team at Oxford University Press, including the referees of this book, for their help in shaping this story; to Tiffany Barry for valuable comments on both words and pictures; and to Andrzej Muszynski, Ian Fairchild, Tiffany Barry, Vicky Ward, Marc Reichow, Yesenia Thibault-Picazo, Grażyna Kryza, and Mike Howe and the British Geological Survey for provision of some of the images used.

List of illustrations

Chapter 1
Primordial rocks

First things first: what is a rock? It is a piece of solid matter that is made up of minerals—a mineral being a chemical compound of fixed composition, or fixed within certain limits. A rock may include a variety of minerals or it can be made up of only one. For instance, a quartzite is a rock that is essentially made up of many grains of the mineral quartz, SiO_2, while the igneous rock anorthosite can wholly consist of crystals of the mineral anorthite, $CaAl_2Si_2O_8$. But if you were given only a single grain of quartz or just one crystal of anorthite you would not really call either of these a rock—rather, they would be a grain and a crystal, respectively.

A liquid—magma, for instance—is not regarded as a rock, but it will turn into one if it cools. This is not just a matter of solidity versus liquidity, but within magma the component particles (atoms and molecules, usually in their electrically charged form as ions—i.e. each atom or molecule either has one or more electrons than it has protons in the nucleus, e.g. Cl^-; or one or more fewer electrons, e.g. Na^+) are moving more or less freely, and have no fixed position relative to each other. Now, if that magma is chilled so strongly that it freezes exceedingly quickly, or if it cools and solidifies as an exceedingly viscous, water-poor melt, these atomic particles will stay randomly dispersed, but now more or less fixed in place. This is solid glass. There are magmatic rocks of this kind—obsidian, formed in rapidly chilled lava, for instance—that are hence amorphous: lacking crystalline structure.

However, if freezing takes a little more time or if the magma is less viscous as it cools, the atoms are able to take up their positions within the molecular lattice of minerals. In these positions, within what is now a crystalline rock, they are in their preferred (lowest) energy state, and so the tendency towards mineral formation in a rock is very pronounced.

The first minerals

If we want to reach back to the earliest minerals of all, we need to take a long journey in space as well as time. For this process, one needs stars—in fact one needs *dying* stars, and so minerals are not part of the primordial matter of the Universe. Rather, minerals appeared some millions of years after the Big Bang, that mysterious and singular event in which all matter and energy (and the laws that govern them) in the Universe originated. The Big Bang, for all of its extraordinary and unrepeated temperatures (briefly, of trillions of degrees), produced only hydrogen (mostly), helium (in modest amounts), and a smattering of lithium: not a promising basis to make minerals from. These primordial elements formed outrushing, expanding, cooling gas clouds. Then, at some stage, gravity came into play.

Gravity pulled parts of these gas clouds together, until they collapsed to form the first stars, and ignited the nuclear furnaces that began transmuting those original elements into those of the rest of the periodic table. The normal processes of nuclear fusion begin this process. But, it is the dramatic finale of large, fast-burning stars—supernovae—that both explosively completes the process of elemental construction and, simultaneously, flings this newly minted matter out into space (Figure 2).

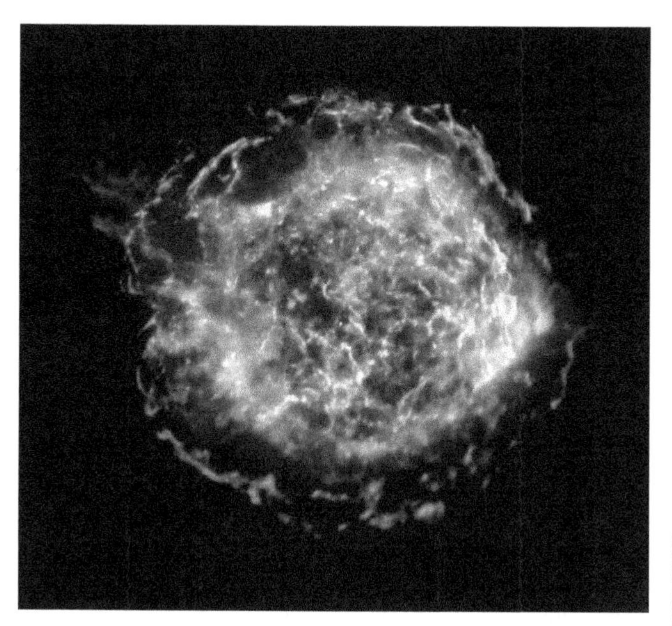

2. Remnant of the Cassiopeia A supernova. It is in stellar explosions like this that most of the chemical elements—the building blocks of rocks—are formed.

The new elements—silicon, oxygen, iron, magnesium, and all the others of the periodic table—initially speed out as high-temperature plasma. As they cool, they condense and then solidify—and the first minerals appear. It is rather a limited array. About a dozen minerals have been detected so far by their spectroscopic signatures, drifting in interstellar dust, or by having been found as 'presolar' grains of interstellar dust within the meteorites (see later) of our own solar system. The presolar grains can be recognized by the patterns of isotopes of their constituent elements (isotopes are forms of atoms with different numbers of neutrons in the nucleus in proportion to the number of protons, which is constant for any given element). Having been formed by different atomic pathways in different supernova explosions, the proportions of different isotopes in these grains can be wildly different to those within our solar system: to a geochemist armed with a mass spectrometer, presolar grains can be as distinctive as oranges within a basket of apples.

These first minerals comprise carbon in the form of diamond, some in tiny whisker-shaped crystals; a few carbides and nitrides; a few oxides, including corundum; and some particles of a form of olivine, which is a member of the most important mineral group (to Earthlings, at least), as a *silicate mineral.*

Silicate minerals dominate the surface of our planet. They are so named because they are based on a combination of oxygen (the next most common element in the cosmos after hydrogen and helium) and silicon. The near-universal building block of these minerals is the *silica tetrahedron*—where each silicon atom is surrounded by four oxygen atoms (Figure 3). The purest form is simply silica, or the mineral quartz, SiO_2, where, to maintain the balance of electrical charges of the ions (Si^{4+} and O^{2-}), oxygen atoms are shared between silicon atoms in the adjacent tetrahedra. In other silicate minerals, other elements can be involved: in olivines, for instance, the tetrahedra form chains, with different combinations of iron and magnesium ions in the structure.

In other silicate minerals, yet other ions can be involved, such as calcium, sodium, and potassium in the feldspar minerals, where the silica tetrahedra form a framework in which aluminium-oxygen octahedra are interspersed. In yet others, the micas, the silica tetrahedra are arranged as sheets, which helps to explain the way in which members of this mineral family split into perfect, thin, parallel flakes along mineral cleavage planes. Other common silicate minerals are pyroxenes, amphiboles, and even the tiny minerals that make up clays. Together, they are the fundamental building blocks of a rocky planet.

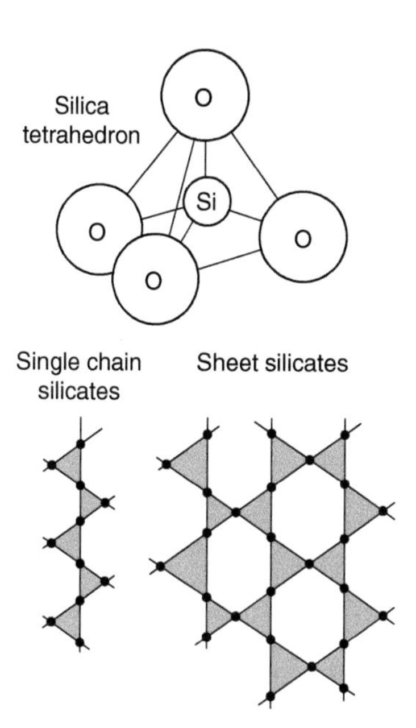

3. The silica tetrahedron—the fundamental component of most crustal rocks—together with a couple of the typical arrangements it takes within minerals.

To return to our primordial mineral formation: perhaps some of these coalesced together, in the supernova aftermath, to create the first microscopic rock fragments. We have no evidence of this, though. For real rocks, a second star generation is needed.

Birth of a star system

The process here is very much like the process that gave rise to the first stars—only here the collapsing gas clouds include that already-formed mineral dust. For the future of everything—including of life itself—this makes all the difference.

As before, most of the collapsing gas-plus-dust cloud aggregates in the central body—the star—which in this case has a more diverse composition than just the original hydrogen, helium, and lithium. Astronomers call these 'metal-rich' stars, to distinguish them from the more primitive 'metal-poor' ones (although by 'metals' in this case astronomers mean any element that is not hydrogen or helium). The processes of real significance to rocks, though, are taking place outside.

The whirling disc of leftover material around the new star includes mineral dust, too, and soon it will become more interesting and diverse. The star lights up, initially through the energy released by the intense compression at its core, and then from fusion, as its nuclear furnaces ignite. The whirling dust near to the star is heated by heat and shock waves emitted by the new sun—and, at least in the case of our own solar system, by its own internal radioactivity—for the mineral material is laced with highly active, short-lived radioisotopes from a nearby supernova explosion. The decay products of these vanished radioactive elements can still be detected in the debris left over from those events.

The mineral material melts and, in places, it clumps together to form droplets which can, in turn, collide and stick together. These are the first rocks known to us as asteroids. They collide, sometimes smashing themselves apart, sometimes aggregating to form planetesimals, kilometres to many kilometres across—the raw materials that would have gone on to build planets. Some of this debris still crashes down on to the Earth today as meteorites, and these exotic chunks of space rock can tell eloquent stories of the birth of a star system.

First rocks in a star system

What kinds of rocks did these primordial processes produce within our own solar system? There is still a good deal of such stuff left over from the planet-building process. Samples of this material still fall out of the sky from time to time—though not as often (or in as large masses) as they would have done in the early days of the Earth. We now recognize them as meteorites, and they provide clues to the times before the Earth existed.

This recognition of meteorites as space debris dates back only a couple of centuries. The rocks themselves, with their fused and melted surfaces, had long been recognized as something special—particularly when they were made of that prized and useful material—native iron—something that is otherwise exceedingly rare at the Earth's surface. In ancient Egypt, iron artefacts from this source—often clumsily made because of the difficulty in achieving temperatures high enough to work this metal—were the preserve of the nobility. A few were found wrapped in Tutankhamun's mummy bandages, including an iron Eye of Horus amulet, to provide protection, royal power, and good health in the afterlife.

These curious iron rocks were debated in classical Greek times. Some philosophers such as Epicurus and Pliny the Elder allowed that rocks might fall out of the sky, while Aristotle considered the heavens incorruptible and flatly rejected the idea that they might be a source of rock debris. Aristotle's views were to dominate Western scientific thought for the next millennium, and so these iron rocks were generally considered to be some atmospheric phenomenon—perhaps 'thunderstones' melted by lightning; or perhaps something to do with volcanoes. The idea that these 'stones' and 'irons' (the word 'meteorite' itself did not enter our language until the early 19th century) could simply fall from the skies was regarded as nonsense, and when in 1794 the German physicist Ernst Florens Chladny linked the half-ton Russian 'Pallas iron' and others with the fireballs occasionally seen in the sky, and proposed that these objects came from outer space, his ideas were widely ridiculed. The following year, however, a 25-kg rock was seen to fall from the sky, landing next to a cottage at Wold Newton, in Yorkshire, where it was discovered, still warm and smoking, deeply embedded in the soil.

The landowner, the colourful Major Edward Topham, exhibited it in Piccadilly. A few years later, in 1803, no less than 3,000 rocks fell out of the sky on to the village of L'Aigle, in Normandy. The young scientist, Jean-Baptiste Biot, sent out to investigate, came to the same conclusion as Chladny, and he gathered enough evidence to convince the doubters. The science of meteoritics was born.

Meteorites today are hunted, not so much among farmlands and forests, but in places where these rare sky-fallen rocks will stand out against their background. The Earth's sand deserts is one such place, and another is the great ice sheet of Antarctica—particularly where the surface ice is slowly turned into water vapour and blown away ('ablated') by dry winds. Deserts are now picked over by local people, who sell the meteorites (often illegally) to collectors, while scientific expeditions are sent out to scour the most promising parts of Antarctica.

Meteorites come in a range of types, but they can be most simply divided into those that are 'stony'—the majority—and those that are 'iron'. Of the stony meteorites, some (*achondrites*) resemble terrestrial igneous rocks of the kind described in Chapter 2. Most, though, when looked at closely, are quite unearthly: these are the *chondrites* (Figure 4).

Chondrites might, at first glance, be mistaken for some kind of sandstone, being largely made of rounded mineral grains. However, these are not the typical sand grains of Earth. Called '*chondrules*', they are melt droplets formed in the violent conditions of the early solar system, before the planets formed, when the dust grains whirling around the just-formed Sun were flash-melted by processes that remain mysterious but that may have included outbursts of solar energy, lightning flashes, and shock waves coursing through the dust clouds. Most recently, it has been suggested that they are melt droplets splashed out from the surface of growing planetesimals as these were bombarded

4. A chondrite meteorite: the particles seen are chondrules, which formed as melt droplets within the early solar system.

by speeding meteorites, making them by-products of planet construction. Some of the chondrules are still made of glass—that is, of a melt that has frozen so quickly that crystals have not had time to form. Others include crystals of iron and magnesium silicates: olivine and pyroxene. In between the chondrules are smaller dust particles, a very few of which have isotopic compositions so extreme that they are regarded as 'presolar grains'—particles of dust that originated in other, more distant, star systems. There are also 'calcium-aluminium inclusions' of very high melting point that seem to be a little older still than the chondrules, and these inclusions have been dated to a very precise 4,566.6 +/– 1.0 million years. This date is held to represent the very beginning of our solar system.

Few of the chondrites that we have are simply pristine samples of stuck-together melt droplets from the dawn of our solar system. Many show evidence of impact shock from collisions with other asteroids. Some have been thermally altered, likely because they originally contained highly active, short-lived radioisotopes, such as ^{26}Al and ^{60}Fe, from a supernova that erupted just before our solar system formed (and that may well have provided the impulse that triggered the gravitational collapse of our parent interstellar dust clouds). In others the original minerals are altered, turned to clay, because the original asteroid material contained water. Even in the depths of space, rocks can be 'weathered'.

One moderately rare but significant type of chondrite is the *carbonaceous chondrite*. These space rocks contain carbon compounds—including some regarded as the building blocks of life, such as the amino acids glycine and alanine. They also often typically show signs of significant water content (such as altered minerals) and the chondrules within them are often small or even absent. Carbonaceous chondrites likely formed farther from the Sun than normal chondrites, where temperatures were lower. As well as being evidence that the organic precursors of life may have had an extraterrestrial origin, their water content has led to suggestions that they may have been a source of a significant part of the water in the Earth's oceans.

Chondrites originated, and broadly remained, as small rocks in space. The achondrites and the iron meteorites mostly represent the shattered remains of larger bodies of rock—large asteroids and planetesimals—that accumulated enough internal heat to begin to melt internally and to separate into a rocky exterior and a metallic core, just as happened on the rocky planets. The melting here was likely helped by the fierce heat provided by those supernova-derived short-lived radioisotopes. The achondrites, other than their fused and melted exterior, can broadly resemble some rocks found at the Earth's surface, particularly those from the Earth's mantle.

Some achondrites have quite specific identifiable sources. There are about a dozen now recognized on Earth that have come from the planet Mars, together with over a hundred that come from the Moon (both identified through their distinctive chemical composition, expressed as particular isotope ratios). Blasted off their parent planetary body by giant meteorite impacts, they drifted through space, and then fell to Earth. This proven means of interplanetary rock transfer has led to some scientists suggesting that the Moon should be searched for rocks from the very early Earth, of four billion years ago and more, for rocks of that age have long been recycled on this geologically active planet of ours.

There is nothing like the iron meteorites at the Earth's surface, because the nearest local source for such rocks, the Earth's core, lies nearly 3,000 km below our feet. This is quite inaccessible to us, and there are and have been no known natural pathways that could ever have brought rock samples from such depths to the Earth's surface. Hence it is through those extraterrestrial 'irons' that we can physically grasp something of the kind of material that lies at the heart of our planet (see Chapter 5). They are mainly made of iron-nickel alloys, the nickel making up from a few percent to a quarter of the content, usually with smaller amounts of cobalt.

The complex iron-nickel crystals reveal beautiful patterns when the cut surfaces of the meteorite are acid-etched. These patterns were first found and described by the obscure English geologist G. Thomson (so obscure, indeed, that no-one now knows what the 'G' of his first name stands for), who discovered them when working in Naples, trying to clean an oxidized iron meteorite specimen. He published on this in 1804, but then his luck ran out quite spectacularly. His correspondence of this to England was lost when the letter-carrier was murdered, while his French-language account, and a subsequent one in Italian, were overlooked by the scientific world of the day. A couple of years later, Napoleon's invasion of Naples forced Thomson to flee to Sicily, where he died shortly after. The structures are now known as Widmanstätten

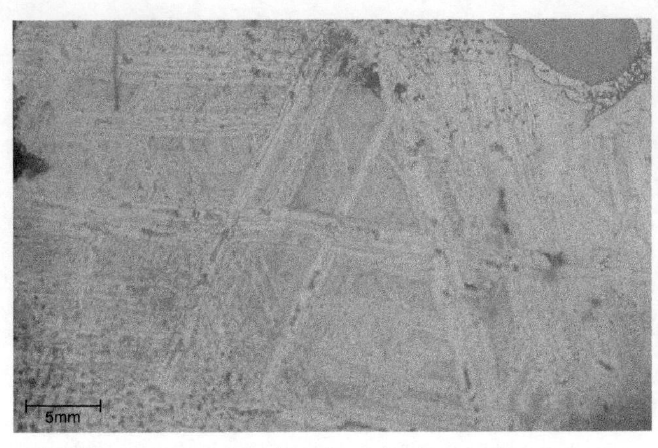

5. A Widmanstätten pattern in an iron-nickel meteorite.

patterns after independent rediscovery by Count Alois von Beck Widmanstätten, the owner of a Vienna porcelain works, who noticed the patterns in 1808 when, out of curiosity, heating iron meteorites (Figure 5).

Widmanstätten patterns reflect the complex formation of large crystals of iron and nickel alloys, over millions of years, within a planetesimal body that was subsequently shattered by colliding with another rock body. They are testament to the complex and violent birth of the nascent solar system.

Among the bodies circling the Sun, the most common rock of all is ice, the main component of comets, and the substance that forms the crust of many of the moons and dwarf planets in the outer reaches of the solar system. Ice as a rock will be discussed later (see Chapter 7). Our next step here is towards the primordial rocks of Earth.

Chapter 2
First rocks on a dead Earth

Violent birth of a new planet

Before Earth, there was Tellus, the name given to the Earth's precursor planet. It was somewhat smaller than the Earth is now, though quite what it was like in its brief existence (of a few tens of millions of years) we do not know and never will. Like every rocky planet or rocky planetesimal large enough to melt internally, iron and nickel and the elements that travelled with them (such as cobalt, sulphur, and antimony) sank to form a metallic core. The remaining lighter elements, dominated by oxygen, silicon, aluminium, together with still-substantial amounts of iron, and magnesium as well as calcium, potassium, sodium, and so on, made up the rocky mantle and crust of this doomed early planet.

Then, in an instant, Tellus disappeared—or, rather, was explosively transformed by collision with a Mars-sized planet that we call 'Theia'. Theia disappeared amid a welter of rock vapour and outflung flash-melted debris. The two cores combined, with Theia's core plunging through the mantle of the stricken Tellus. Much of the rock vapour and debris coalesced to form the Moon. The suddenly born Earth, now enlarged by much of the matter of what used to be Theia, was a glowing, rapidly spinning globe. Just after the impact, this new planet had a surface magma ocean, perhaps thousands of kilometres deep. And now the stage was set for many things—including the development of the most diverse, beautiful, and informative rocks in the solar system.

To make a variety of rocks, there needs to be a variety of minerals. The Earth has shown a capacity for making an increasing variety of minerals throughout its existence. Life has helped in this—but that part of the story comes a little later. Even a dead planet—which the Earth originally was—can evolve a fine array of minerals and rocks. This is done simply by stretching out the composition of the original homogeneous magma.

Primordial rocks of Earth?

Such stretching of composition would have happened as the magma ocean of the earliest, post-Tellus, Earth cooled and began to solidify at the surface, forming the first crust of this new planet—and the starting point, one might say, of our planet's rock cycle. When magma cools sufficiently to start to solidify, the first crystals that form do not have the same composition as the overall magma. In a magma of 'primordial Earth' type, the first common mineral to form was probably olivine, an iron-and-magnesium-rich silicate. This is a dense mineral, and so it tends to sink. As a consequence the remaining magma becomes richer in elements such as calcium and aluminium. From this, at temperatures of around 1,000°C, the mineral plagioclase feldspar would then crystallize, in a calcium-rich variety termed anorthite. This mineral, being significantly less dense than olivine, would tend to rise to the top of the cooling magma. On the Moon, itself cooling and solidifying after its fiery birth, layers of anorthite crystals several kilometres thick built up as the rock—anorthosite—of that body's primordial crust. This anorthosite now forms the Moon's ancient highlands, subsequently pulverized by countless meteorite impacts. This rock type can be found on Earth, too, particularly within ancient terrains.

Was the Earth's first surface rock also anorthosite? Probably—but we do not know for sure, as the Earth, a thoroughly active planet throughout its existence, has consumed and obliterated nearly all of the crust that formed in the first several hundred million years of its existence, in a mysterious interval of time that we now call the Hadean Eon. More recent anorthosites have, however, formed by somewhat different processes, with some kind of physical crystal segregation within a magma being suggested for many examples.

The earliest rocks that we know of date from the succeeding Archean Eon. It was likely midway through the Archean Eon, perhaps some three billion years ago, that a planetary mechanism started up that was to determine the nature of all subsequent rocks on Earth. This was *plate tectonics*, a mechanism that is, as far as we know, unique to our planet.

It was glimpsed from early in the 20th century by the pioneering (and, at the time, widely derided) German scientist Alfred Wegener, who proposed, based on a variety of evidence (most good; some mistaken), that continents had drifted relative to each other by thousands of kilometres through geological time. A little later, the British geologist Arthur Holmes surmised that such drift might be driven by slow-moving convection currents in the mantle. The idea took decades to be accepted, but plate tectonics is now the ruling—and extraordinarily successful—paradigm of the Earth sciences.

In plate tectonics, the Earth's surface is divided into rigid rocky plates, about a couple of hundred kilometres thick, which make up what is termed the *lithosphere*. These plates move across the Earth's surface, being separated from the deep rock of the Earth's mantle below by a weak and deformable layer called the *asthenosphere*. The plates may diverge, converge, or slide past each other. Where they diverge is a site of large-scale rock production, a remarkably uniform production line that creates the ocean crust. A key process here is fractional melting.

Fractional melting and basalt

The starting point for most igneous rocks today is not simply from an enormous pool of magma (as was the case with the post-Theia magma ocean). Rather, it is a mass of solid rock, which is what most of the Earth is now made of, after more than four billion years of cooling following the Theia impact. If this rock is tens of kilometres underground, and hot enough to be *almost* melting, then conditions may be created such that it *just* begins to melt. This need not involve an increase in temperature, because pressure is a factor too: the higher the pressure, the greater the tendency for rock to remain solid.

Where plates are pulled apart, then pressure is released at depth, above the ever-opening tectonic rift, for instance beneath the mid-ocean ridge that runs down the centre of the Atlantic Ocean. It was the oceanographers Bruce Heezen and Marie Tharp who, in the 1960s, recognized that the mid-ocean ridges are not just enormous underwater mountain ranges (for mountain ranges are associated with compression). Rather, they possess, running along their centre, a continuous rift valley—a sign that the oceanic crust here was in reality stretching and cracking apart (Figure 6).

The pressure release from this crustal stretching triggers *decompression melting* in the rocks at depth. These deep rocks—peridotite—are dense, being rich in the iron- and magnesium-bearing mineral olivine. Heated to the point at which melting *just* begins, so that the melt fraction makes up only a few percentage points of the total, those melt droplets are enriched in silica and aluminium relative to the original peridotite. The melt will have a composition such that, when it cools and crystallizes, it will largely be made up of crystals of plagioclase feldspar together with pyroxene. Add a little more silica and quartz begins to appear. With less silica, olivine crystallizes instead of quartz.

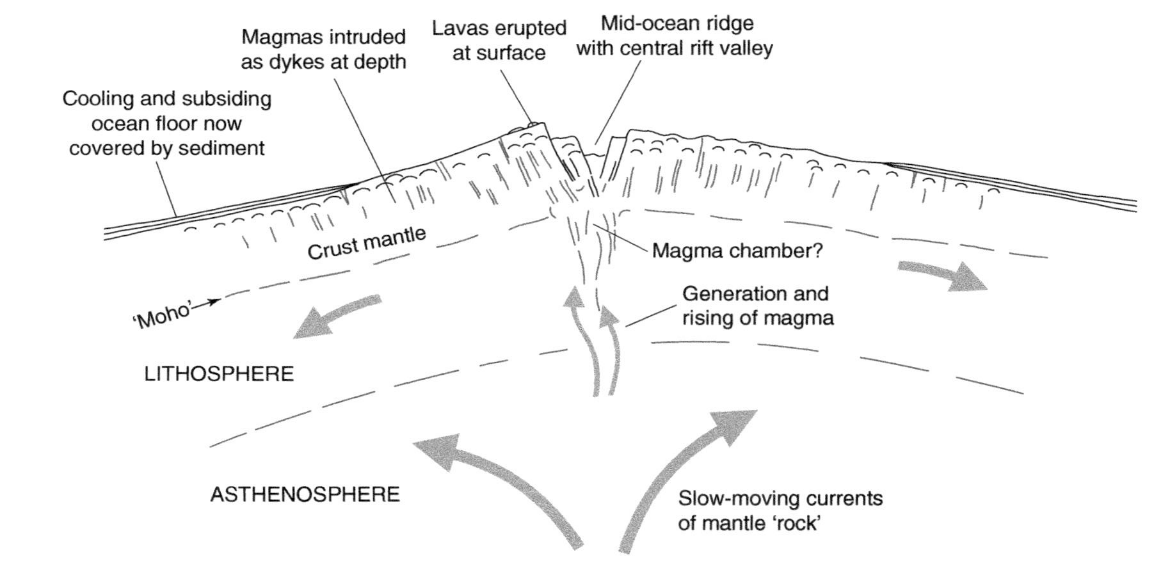

6. Diagram of a mid-ocean ridge cross-section to show magma production.

The resulting rock is basalt. If there was anything like a universal rock of rocky planet surfaces, it is basalt. On Earth it makes up almost all of the ocean floor bedrock—in other words, the ocean *crust*, that is, the surface layer, some 10 km thick. Below, there is a boundary called the *Mohorovičič Discontinuity* (or 'Moho' for short), named after the Croatian seismologist Andrija Mohorovičič. The Moho separates the crust from the dense peridotitic mantle rock that makes up the bulk of the lithosphere.

More basalt erupts from 'hot spots' beneath the crust, some almost certainly related to ascending fountains of mantle material, termed *plumes* (see Chapter 5). The classic example of this is the Hawai'ian island chain, 5,800 km long, that, over eighty-five million years, formed as the Pacific Ocean plate passed over the hot spot. Each island formed as a basaltic volcano above the hot spot, eventually building up to a height of 4 km to reach the ocean surface, and then a further 1 km or more, before becoming extinct as it passed beyond the hot spot. The currently active island, Hawai'i, or 'Big Island', includes Mauna Loa, the world's biggest single volcano, which produces classic examples of basaltic volcanism.

There are larger masses of basalt too: oceanic plateaux, such as the enormous, mostly submerged, Ontong-Java plateau, that represent relatively brief episodes of more enhanced basaltic volcanism, again possibly due to the head of a rising plume impinging on the base of the lithosphere. Related phenomena—large igneous provinces representing vast outpourings of magma—also occur on the land of the continents. The Giant's Causeway in Northern Ireland forms part of one such phenomenon. Others are the Deccan Traps hills of India, the stark topography of the Snake River valley in Idaho, and the Faroe Islands that formed as the Atlantic Ocean opened (Figure 7).

By-products of the accelerated production of this rock type—loading of the atmosphere with carbon dioxide, sulphur

7. Basalt 'trap' terrain: a succession of basalt lavas from the opening of the North Atlantic, fifty-five million years ago: Faroe Islands.

dioxide, chlorine, and fluorine—have been linked with environmental crises, notably the strong correlation between the enormous 'Siberian traps' basalts which erupted in a brief and savage burst of volcanism 252 million years ago, and the greatest mass extinction event of all, at the boundary between the Permian and Triassic periods.

Basalt makes up most of the surface of Venus, Mercury, and Mars (see Chapter 7). On the Moon, the 'mare' ('seas') are not of water but of basalt. Basalt, or something like it, will certainly be present in large amounts on the surfaces of rocky exoplanets, once we are able to bring them into close enough focus to work out their geology.

Cooling rate and crystal size

Basalt is the result of basaltic magma that has flowed onto the Earth's surface, and so has cooled and frozen rapidly. When this happens, there is little time for the atoms and molecules in the magma to make their way, by diffusion, to the surfaces of growing crystals. Hence, basalt is made of a mass of tiny crystals, which can only be really seen using a microscope—geologists say it is 'fine-grained'. If cooling is exceptionally rapid (as at the chilled surface of basaltic magma that has come into contact with water) then there is no time at all for crystals to grow, and volcanic glass forms instead. Such glass is inherently unstable, or more precisely *metastable*, because the atoms and molecules are in higher energy states than they would be if they were locked into the ordered molecular structures of crystals. Hence, over millions of years—and even in the solid state—the atoms and molecules diffuse, very slowly, the tiny distances necessary to lock onto other atoms and molecules, and they begin to form incipient crystals. The rock thus *devitrifies* and turns from clear to opaque, as a myriad submicroscopic crystal boundaries scatter the light.

Volcanic glass is much sought after by geologists working on ancient magmas—precisely because it represents almost perfectly samples of those long-cooled magmas. It can be preserved as minute droplets, for instance, preserved within growing crystals, and these can be analysed by finely focused electron beams linked to spectrometers to work out the composition of the original melt material.

If, though, basaltic magmas cool underground within their magma chambers, insulated by the surrounding rocks, they can cool very slowly, over many thousands of years. This gives time for large crystals to grow, which can be several millimetres or even a few centimetres across, forming an igneous rock known as a gabbro. Smaller bodies of magma, such as those intruded as sheets into fractures in the rock, cool at moderate rates to give the medium-grained dolerites. Dolerites are common: they form the 'bluestones' of Stonehenge, for example, and the impressive Palisades cliffs of the River Hudson by New York City.

It is common for magma to begin cooling slowly underground and then, with crystallization only partly complete, to be suddenly erupted onto the surface because of some change in a volcano's internal plumbing. The resultant rock is termed a *porphyry*, and includes large *phenocrysts*—early, slow-grown crystals set within a rapidly chilled and hence finely crystalline groundmass.

One can see these large crystals with the naked eye, and examine them more closely with a magnifying glass, especially in the form of the small hand lens—a form of technology that is still indispensable to any field geologist. But to see them in beautiful detail, one needs to cut a sliver of rock and stick it onto a glass slide, then grind it down until it is translucent, just a thousandth of an inch thick. This is a *thin section* of rock that, examined in polarized light using an optical microscope, can reveal extraordinary subtleties of rock texture and mineral composition.

The making and analysing of rock thin sections was pioneered in the mid-19th century by an ingenious amateur scientist Henry Clifton Sorby, who lived and worked in Sheffield (he had inherited a private income, so spent his life energetically and productively following his scientific interests). Sorby said that using this technique he could indeed (to turn the tables on his early detractors) 'examine mountains with a microscope'. These days, a geologist's armoury is extended by such sophisticated equipment as scanning electron microscopes and electron microprobes, but the optical examination of thin sections remains fundamental to understanding rocks (Figure 8).

Magma evolution

Basaltic magmas are a common starting point for many other kinds of igneous rocks, through the mechanism of fractional crystallization described earlier. Remove the early-formed crystals from the melt, and the remaining melt will evolve chemically, usually in the direction of increasing proportions of silica and

8. A porphyritic basalt seen in microscope thin section showing large phenocrysts (formed during slow cooling) set within a finer grained groundmass (formed during later rapid cooling).

aluminium, and decreasing amounts of iron and magnesium. These magmas will therefore produce *intermediate* rocks such as andesites and diorites in the finely and coarsely crystalline varieties, respectively; and then more evolved silica-rich rocks such as rhyolites (fine), micrograpnites (medium), and granites (coarse). Such silica-rich rocks are often referred to, especially in older literature, as 'acid' igneous rocks as opposed to the 'basic' igneous rocks such as basalt—though it is now more common to refer to them as *felsic* and *mafic* igneous rocks, respectively.

In yet older literature, dating back to the beginnings of geology, there was a school of thought of the 'neptunists' who argued that that granites and allied rocks had crystallized from seawater onto the sea floor, much as crystals of rock salt can crystallize out from a concentrated brine (see Chapter 3). The best-known advocate of neptunism was Abraham Gottlob Werner, who in the late 18th century was an influential professor of mineralogy at Freiburg in Saxony. One of his supporters was the poet Goethe (himself once a superintendent of mines, with the iron mineral goethite named after him). Goethe, in his epic poem *Faust*, portrays a dialogue between a neptunist and a member of the opposing school of plutonism, arguing for a fiery origin for such rocks—and Goethe's advocate of plutonism was no less than Mephistopheles, the devil. It did not take long, in reality, for the diabolical plutonists, who included such early savants of geology as James Hutton and Charles Lyell, to win the argument.

Granites themselves can evolve a little further, especially at the late stages of crystallization of large bodies of granite magma. The final magmas are often water-rich ones that contain many of the *incompatible* elements (such as thorium, uranium, and lithium), so called because they are difficult to fit within the molecular frameworks of the common igneous minerals. From these final 'sweated-out' magmas there can crystallize a coarsely crystalline rock known as pegmatite—famous because it contains a wide variety of minerals (of the ~4,500 minerals officially recognized on Earth—see Chapter 8—some 500 have been recognized in pegmatites). Including tourmaline, topaz, and beryl, these minerals may be of gem quality and can be a source of lithium, tin, molybdenum, and other rare elements, including the 'rare earth' elements such as neodymium and dysprosium, now so important to our high-technology industries.

Such fractional crystallization processes to form more silica-rich rocks can take place as the ocean crust forms. However, such rocks are more common in that other great foundry of magmatic rocks on Earth—the zones where tectonic plates collide. If basalt is the signature rock for crustal expansion and mantle hotspots, then it is the destruction of basaltic ocean crust that leads to the formation of a different—and geologically more long-lasting—suite of rocks: that of the continental crust.

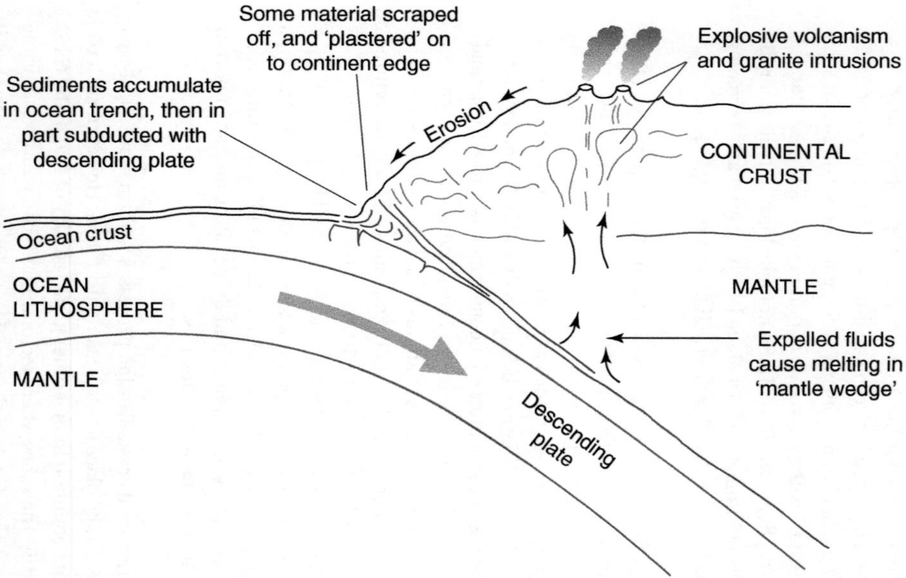

9. Diagram of rock formation at a subduction zone.

Where tectonic plates collide

At any one time, ocean floor basalts are the most common rock type on our planet's surface. But any individual piece of ocean floor is, geologically, only temporary. It is the fate of almost all ocean crust—islands, plateaux, and all—to be destroyed within ocean trenches, sliding down into the Earth along subduction zones, to be recycled within the mantle. From that destruction, though, there arise the rocks that make up the most durable component of the Earth's surface: the continents.

In any subduction zone, the descending oceanic plate is a slab of rock that is not simply the crust, but is the larger mass that is termed the lithosphere (discussed earlier in this chapter). This (the crust together with the uppermost, rigid part of the mantle) decouples from the underlying mantle along the weak, melt-prone asthenosphere.

The descending lithosphere is cold and, in its upper levels, water-soaked. As it descends, much of the water is forced out, initially by simply squeezing the rock mass and then by the transformation of minerals from hydrous to anhydrous forms—a process that may take place suddenly enough to generate earthquakes. At depths down to 300 km, a good part of the remaining water migrates upwards, much of it into the overlying wedge of mantle rock. This addition of water promotes melting in that mantle wedge (even despite some cooling caused by proximity with the descending cold oceanic crust). The melt produced in this fashion is not typical basaltic melt, but is more silica-rich, typically being andesitic ranging to rhyolitic (with even more silica). The tendency to become silica-rich is strengthened by another process that takes place—the melting of other rocks in the crust by the hot magma; and the assimilation of this newly molten material into the rising magma.

This magma ascends to feed the characteristic volcanoes of the mountain belts and island arcs that overlie subduction zones (Figure 9). These are not the gently simmering basaltic volcanoes, such as those on Hawai'i. The silica-rich, water-rich magmas behave very differently, and produce rocks that are different in texture as well as in the minerals that make them up. These magmas are much more viscous, partly because of a much greater degree of polymerization of the silica tetrahedra within the melt. Such lavas, when they erupt, flow much more slowly and stiffly than do the fast-flowing and fountaining basaltic lavas.

Most of the magmas do not make it to the surface as coherent lavas. As they rise towards the surface, the pressure on them lessens, and the gases dissolved within them begin to form bubbles and try to escape, just as bubbles form in a bottle of fizzy drink when the cap is removed. In the fizzy drink—and also in a runny basaltic magma—the bubbles can escape gently and easily. Not so in these silica-rich melts. The rock becomes a rapidly expanding viscous magma foam (pumice) that can then explode into an inferno of shattered bubbles and crystals, entraining within it rock fragments from the shattered volcanic vent and crater. Part is then carried kilometres up into the sky in an eruption column by the formidable heat engine of the volcano, to fall out of the eruption cloud as volcanic ash. If the mixture is too dense to be carried aloft, it spills over the volcano as a fearsome, rapidly-moving, ground-hugging *pyroclastic current*, which is lethal to any living thing in its path. The ash that has fallen out of the air drapes the entire topography around, thinning away from the volcano, while the deposits of the pyroclastic currents (called *ignimbrites*) accumulate in valleys and other low-lying areas.

The rocks produced are distinctive. The volcanic ash that has fallen out of the eruption column (technically termed *airfall tuff*) largely comprises fragments of pale, silica-rich pumice. If such material lands in the sea, it floats until it either becomes waterlogged enough to sink or it washes up onto a shore. It includes *lithics*: denser fragments of volcanic rock that have generally broken off the crater walls. In a large eruption, fist-sized fragments of such rock can be carried more than 10 km into the air in the hot, turbulent updraught from the volcano, before they eventually fall out from the spreading eruption cloud.

Time-keeper rocks

Igneous rocks of these kinds have been produced as long as the Earth has been in existence, and our direct record of them goes back some four billion years (with indirect evidence in the form of individual crystals scavenged from yet older rocks dating to 4.4 billion years). Our ability to date these rocks, and so to place them in a time order and thus make sense of Earth's history, is due to the radioactive character of some minerals. Many felsic igneous rocks contain small amounts of minerals such as zircon (zirconium silicate) and monazite (a complex phosphate of rare earth elements).

Both of these minerals, when they form, can accommodate some radioactive uranium (the parent element) within their crystal lattice. Over time, the uranium decays to form a chain of daughter elements at the end of which is the element lead. The rate of decay is impervious to changes in heat, pressure, or chemical environment, and it is referred to in terms of its half-life—the time taken for half of the parent element to decay away (e.g. this is 704 million years for uranium-235 decaying to lead). The reliability of this radiometric dating method (or 'atomic clock') depends on the original crystal retaining both the parent and daughter elements within it (and not absorbing more of either from the external environment). With modern analytical techniques (most notably a mass spectrometer, which can analyse levels of parent and daughter elements in tiny amounts and to high precision), the errors on dates can be well under 1 per cent of the age of the mineral.

There are now quite a number of such atomic clocks. The mineral sanidine, for instance (a kind of potassium-rich feldspar), includes a radioactive isotope of potassium that decays into the gas argon. There are other radiometric dating methods based on the decay of a radioactive isotope of rubidium to strontium; of samarium to neodymium; and so on (the most widely known of such methods, using carbon-14, involves too short a half-life to be of common use in most rocks, but it has been used, say for dating wood fragments caught up in sedimentary deposits that are less than about 60,000 years old, this being the limit of this technique). By and large, acid rocks, which are more highly fractionated, contain more dateable minerals than do basic rocks, but even the latter can now commonly be dated, by seeking out the small amounts of radioactive minerals that they contain.

The ability to date igneous rocks by their radioactive content is one means of placing the rocks of the Earth within a time context, and so one key to deciphering the history of our planet. Another means of establishing history from rocks is geometrical, not chemical, and it relates to the next major group of rocks that we shall discuss. These are the strata of the sedimentary rocks.

Chapter 3
Sedimentary rocks

The Earth is a unique planet in the solar system, in many ways. It is the only planet with any form of life (as far as we know); the only one with abundant liquid water at the surface (Mars can occasionally accumulate tiny drops of concentrated brine dew); and it is the only planet that releases its internal heat via the remarkable mechanism of plate tectonics. All this contributes to its having the greatest abundance and variety of sedimentary strata in our star system, their history stretching back some 3.8 billion years. And, from what little we know of planets beyond our own star system, the Earth's sedimentary riches are probably a cosmic rarity.

The smooth running of plate tectonics is the factor that continually lifts regions of crust high into the atmosphere, as mountain belts, to be exposed to erosion and weathering. The rocks, fractured and jointed from the tectonic forces that acted on them underground, are subject to gravity, intermittently tumbling from the mountains in landslides and avalanches. The resulting debris is then set to undergo protracted transformations, in the next stage of the Earth's rock cycle, to take the journeys that will change it into different types of sedimentary strata.

Rain and river-water, that product of the Earth's active hydrological cycle, act on the rocks in two ways. In cold climates, daily or seasonal freezing and thawing of water trapped in surface rock acts to open up fractures and further weaken the rock mass—physical weathering. Then there is the chemical action of the water, particularly as it is generally weakly acidic from dissolved atmospheric carbon dioxide and humic substances in the soil. When coming into contact with the minerals of rocks such as granites and basalts—forged at high temperatures and often at high pressures, and therefore unstable in cool surface conditions—it dismantles their molecular frameworks. In general, the higher the temperatures at which the minerals formed, the more they are prone to this chemical attack. Olivines, pyroxenes, micas, feldspars, and other high-temperature minerals are broken down, and the rocks containing them become soft, crumbly, and easily blown or washed away.

A part of the mineral substance goes into solution, and is washed into rivers, ultimately to add to the salt content of the sea. Of the rest, part is transformed into a new family of minerals—the clay minerals—that, as we will see, play key roles in many processes at the Earth's surface, both biological and chemical; and part remains more or less unaltered. These unaltered fragments include large rock fragments—boulders, cobbles, and pebbles—and particles of more stable minerals, chief among them being quartz.

These sedimentary particles are then transported further by gravity, wind, and water, until they reach the basin where they accumulate. The mode of transport is central to determining the nature of the sediment, and hence of the resulting sedimentary rock. Key to this is the process of sorting. Some forms of transport result in very little sorting of the original mass of sediment. These include debris flows and the 'boulder clays' (more properly now termed *glacial tills*) transported beneath moving ice-sheets, where grains of all sizes are mixed and travel together. Flows of wind and water, though, will separate grains into different size classes (more specifically, of aero- or hydrodynamic equivalence) (Figure 10).

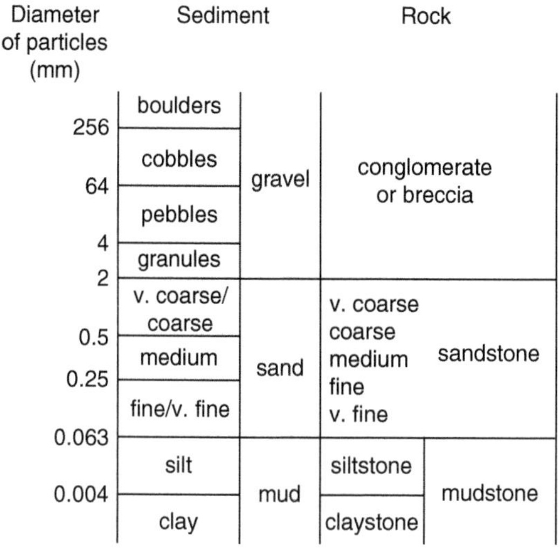

Diameter of particles (mm)	Sediment		Rock	
	boulders	gravel	conglomerate or breccia	
256				
	cobbles	gravel	conglomerate or breccia	
64				
	pebbles	gravel	conglomerate or breccia	
4				
	granules	gravel	conglomerate or breccia	
2				
	v. coarse/ coarse	sand	v. coarse coarse	sandstone
0.5				
	medium	sand	medium	sandstone
0.25				
	fine/v. fine	sand	fine v. fine	sandstone
0.063				
	silt	mud	siltstone	mudstone
0.004				
	clay	mud	claystone	mudstone

10. The standard grain size categories, and equivalent sediment and rock names for sedimentary rocks.

Thus, rock fragments too heavy to transport by, for instance, a river current of moderate strength, may be left behind to pile up as pebble- and cobble-rich gravel that will only be swept along the river channel floor during floods. Eventually such cobble and pebble deposits may be swept—now rounded and diminished in size from many collisions with other rock fragments—onto a pebbly beach, where further rounding will take place through the action of the waves. Sometimes, pebbles from a beach can be storm-swept offshore, perhaps to pile up as a gravelly layer on the bottom of some submarine canyon. In each of these environments, the gravel might eventually be fossilized, becoming a hard rock (see later), in which form it is known as a conglomerate (Figure 11) or, if the fragments still retain their original angular edges, a breccia.

Sand may be swept as an intermittently mobile carpet along the river floor, while silt and mud will be carried far in suspension, to

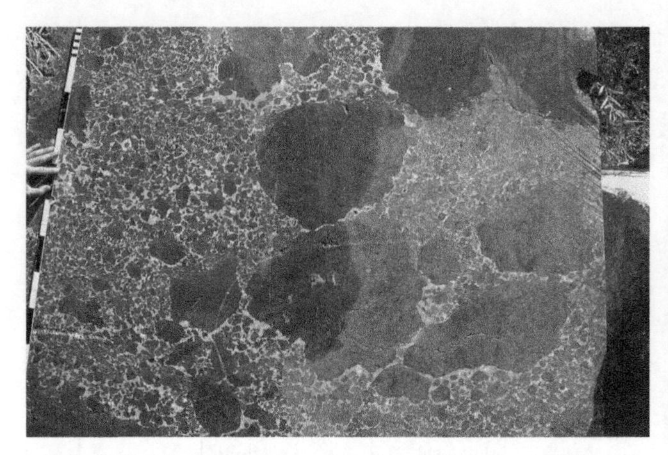

11. The 'Scheck breccia' of Adnet, Austria, a rock made of pebbles, cobbles, and boulders of dark red limestone cemented by pale calcite.

settle in areas of quieter water—perhaps a lake or sea that the river has flowed into—or more locally mud can drape an area of floodplain following a flood when a river has burst its banks.

Wind is particularly good at winnowing and sorting sediment, producing the great sand seas of arid deserts (and, in gentler climes, coastal sand dunes along shorelines). The grains here are typically very well rounded in addition to being well sorted, a result of the countless mid-air collisions with other sand grains, the impacts producing a characteristic frosted texture to the grain surfaces.

Sand shapes

Among the most common—yet in some ways still the most beautiful and enigmatic—phenomena associated with the transportation and sorting of sediment by currents of wind and water are the variety of geometrical shapes assumed by the moving sediment, notably ripples and dunes.

The succession of these is reasonably well understood, particularly for sand-sized sediment. A gentle flow of water will first begin to move sediment as single-grain carpets, which produce a horizontal lamination in sediment deposited in such conditions. Increase the current speed, and the sediment builds into small regular shapes—moving sand ripples, which at first have a straight crestline. Then, as current speed increases and the flow becomes more turbulent, the ripples become more wavy in shape when viewed from above, before breaking up into individual crescent- or diamond-shaped structures. As the current increases still further, much larger structures form, now decimetres or even metres in height. These are dunes, and geometrically resemble current ripples, not least in also having both straight-crested to wavy-crested and isolated forms.

Both ripples and dunes commonly leave their mark in sedimentary rocks. This can be seen as the three-dimensional ripple or dune form on the surface of a stratum. More commonly, though, broken or cut sections through such a stratum will reveal inclined layering that represents successive positions of the moving front of the ripple or dune. This is termed *cross-stratification* (or *cross-lamination* in the case of sections through ripple-forms), and can be used to work out the directions from which ancient currents flowed (Figure 12).

At yet higher velocities, the flood of water washes out the dunes, leaving a flat bed of sand that may preserve horizontal lamination, which can be distinguished from the lamination formed in very gentle currents by long, but very thin scours created in the sediment by fast-travelling vortices. These parallel streaks (termed *primary current lamination*) may often be seen in sandstone pavement slabs, and betray the high-flow conditions in which the original sand was laid down.

Yet higher flow speeds—of the kind attained by catastrophic floods racing down mountainsides or within that rapid high-temperature

12. Sandstone with cross-stratification representing the migration of sand dunes driven by river currents about 360 million years ago. (Portishead, England.)

current that is a pyroclastic current pouring out of an erupting volcano—result in another kind of dune, termed an *antidune.* These are gently inclined structures that often build and migrate *backwards* (by accumulation of sedimentary particles on their upslope face) even as the rest of the speeding particles stream downhill.

Sedimentary structures in rock strata can reflect more specific processes than just current speed. The swash and backwash of waves along a coastline produces layers of sand and pebbles inclined at a gentle angle to the horizontal, and these can be fossilized, to betray the presence of an ancient beach. In water at paddling depth, the waves create little symmetrical, sharp-crested ripples, while further out to sea pounding storm waves can pile up sand and shell fragments into a pattern of low hummocks, a metre or two apart, on the sea floor. The resultant *hummocky cross-stratification* developed in sandstones is such a good guide to prehistoric tempests that it has its own acronym—'HCS'. The action of tides, too, can leave its mark in the way that ripples and dunes are swept first one way by the flood tide and then in the opposing direction during the ebb, and by the way mud can settle out as thin layers in slack water at high or low tide.

Ripples and dunes are also built by currents of air. Sand dunes are in some ways similar to their counterparts formed in water, and indeed fossilized examples can be hard to tell apart from those formed in water. Wind ripples, though, while superficially resembling water-formed current ripples, are entirely different in structure. They are effectively slow-moving small barricades of the coarsest grains that accumulate on the ripple crests, and when fossilized appear as planar structures (called *pinstripe lamination*) rather than as the cross-lamination of water-formed ripples. The action of wind on sand has been studied by many, but perhaps none more remarkable than that scientifically minded brigadier, Ralph Alger Bagnold.

Bagnold fought in both World Wars, in the second one founding and leading the British Army's Long Range Desert Group (a precursor of the SAS). Fascinated by deserts, he not only explored them between the wars—making the first recorded east–west crossing of the Libyan Desert, for instance—but also pioneered a rigorous, mathematically based study of the desert sands. His 1941 book, *The Physics of Blown Sands and Desert Dunes*, is still regarded as a classic: indeed a still-functional classic, as it continues to be cited in NASA's studies of the wind-blown dunes of the planet Mars, where a dune field is named after him, and of Saturn's frigid moon, Titan (see Chapter 7).

Ripples and even dunes may be recreated in the laboratory, where the flow conditions that form them can be analysed. But it is still unclear as to why they take the shape they do, or what function they have (if indeed one can speak of a function for such structures). They tend now to be thought of as 'self-organizing structures' with 'emergent properties' that clearly reflect not just the sand's grain characteristics but also the dynamics of the entire flowing air- or water-grain system. Whatever the subtleties underlying their formation, they remain among the most common, and commonly studied, features of sedimentary rocks.

Mud

Mud is the commonest sediment on our planet's surface, and the rock that is made of it, *mudrock* (also called *mudstone* or *shale*) is the commonest sedimentary rock on the planet by far. It is a very Earthly sediment; there is a little mud, but not much, on Mars, probably none to speak of on Venus, and probably only small amounts on other rocky bodies in the solar system. Although common on Earth, it is also marvellously potent stuff, with great significance for life on Earth and, currently, great significance too for the maintenance of our modern industrialized society.

Mud can settle anywhere from the bottom of a puddle or a pond to accumulating on great expanses of the sea floor. It is made up of various mixtures of clay minerals, clay- and silt-sized grains of quartz, and other minerals, often with a fair admixture of organic material, both dead (decayed and degraded remains of animals and plants) and alive (at the surface of the sediment there are thriving microbial populations, in part feasting on those remains). These components are a good deal more complex than are the rounded mineral particles of sand.

Clay itself is a remarkable material, comprising a family of sheet-like minerals, comparable to micas but microscopic in size. Appearing book-, flake-, or strand-like when viewed by electron microscope, clay minerals have enormous surface areas relative to their mass—up to several hundred square metres per gram. At that scale of mineral size, patterns of electrical charge become important, so that clay mineral surfaces are often negatively

13. Finely laminated mudstone (Jurassic; Port Mulgrave, Yorkshire, England). The pale object to the right of the hand lens is a calcareous concretion, which has grown underground by chemical precipitation.

charged. In fresh water, individual flakes thus repel each other, while in seawater the abundant, charged ions neutralize this repulsion, so that the clay flakes clump together ('flocculate') and settle more easily onto the sea floor.

These variable charge patterns mean that clay flakes can absorb or desorb other ions—or water molecules—more or less loosely onto the surface, thus acting as ion exchangers. Humans exploit these qualities with the clay mineral montmorillonite, in fuller's earth, for de-greasing, and in facial mud-packs; and we are subject to the same qualities when strata containing this montmorillonite expand in winter, as they absorb water, and shrink in summer, on drying, thus disturbing building foundations.

Orders of magnitude more complex still than the protean clay minerals are the various forms of organic matter that collect in mud. A spadeful of mud from a pond or a tidal flat shows the appearance—and smell—of this organic matter. It is a dynamic mixture of living and dead organisms ranging from microbes to worms, molluscs, and crustaceans: a tiny sample of a worldwide ecosystem. Much of the organic matter is respired—oxidized—but some is buried, to join the giant organic carbon store held within rock strata (Figure 13). A key factor in determining whether or not organic matter is buried together with the sediment grains is the mud's oxygenation level, for oxygen is soon used up in such a setting, and anoxic muds, in which air-breathing organisms cannot survive, are usually less than a spade's depth below the surface.

This is the gateway to the Earth's chemical underworld, where both living and mineral assemblages are transformed. Life here comprises anaerobic microbes that, in the absence of free oxygen, strip the oxygen atoms from dissolved sulphate ions for energy, leaving sulphide ions that combine with reduced iron to produce tiny crystals and coatings of pyrite. These can leave the sheen of 'fool's gold' on even recently dead and buried shells. Lower down, the microbes soon use up the sulphate and begin to ferment the organic matter, releasing methane gas. They can carry on, in ever smaller number and ever more slowly, eking out a living on progressively altered and indigestible organic fragments for 1 km or more below the surface.

The less oxygen there is, the more the organic matter is preserved into the rock record, and it is where the seawater itself, by the sea floor, has little or no oxygen that some of the great carbon stores form. As animals cannot live in these conditions, organic-rich mud can accumulate quietly and undisturbed, layer by layer, here and there entombing the skeleton of some larger planktonic organism that has fallen in from the sunlit, oxygenated waters high above. It is these kinds of sediments that will go on, later in their history (see later), to generate the oil and gas that currently power our civilization.

Limestones

Of the dissolved ions that rivers carry down to the sea, some can accumulate for tens of millions of years in the seawater—such as the highly soluble sodium and chloride that form much of the salt in the ocean (these may eventually be deposited as layers of rock salt, as seas dry up—see later). Others, less soluble, quickly come out of solution. Today, this includes iron (although, as we shall see, it was not always so), and some rare elements such as neodymium (used because of this as a 'tracer' of past seawater movements when preserved in rocks). In between, there are ions of calcium, magnesium, and carbonate, present in moderate amounts in seawater, but also mostly at saturation (or even at supersaturation) levels. These ions are ripe for taking out of water as the components of limestone. Limestone is made of the minerals calcite and aragonite—two crystallographically different forms of calcium carbonate; and, in places, of *dolomite*, the term used both for a mineral—calcium/magnesium carbonate—and for the rock made of this mineral.

Calcium carbonate, being durable and protective, is the mineral of choice for many organisms, both plants and animals, for building their skeletons. The plants range from single-celled plankton to bottom-living large algae, while the animals include corals and a multitude of shelled animals (see Chapter 6).

The rock that results from the accumulated remains of these organisms is often spectacularly fossiliferous (Figure 14). However, limestones would form even on a dead Earth, for the calcium carbonate would still have to crystallize out of solution once its saturation levels became too great. This kind of crystallization indeed takes place today on a large scale in the oceans. Around islands such as the Bahamas, where the waters are warm (calcium carbonate is one of those few minerals which becomes *less* soluble as the water warms) and the climate

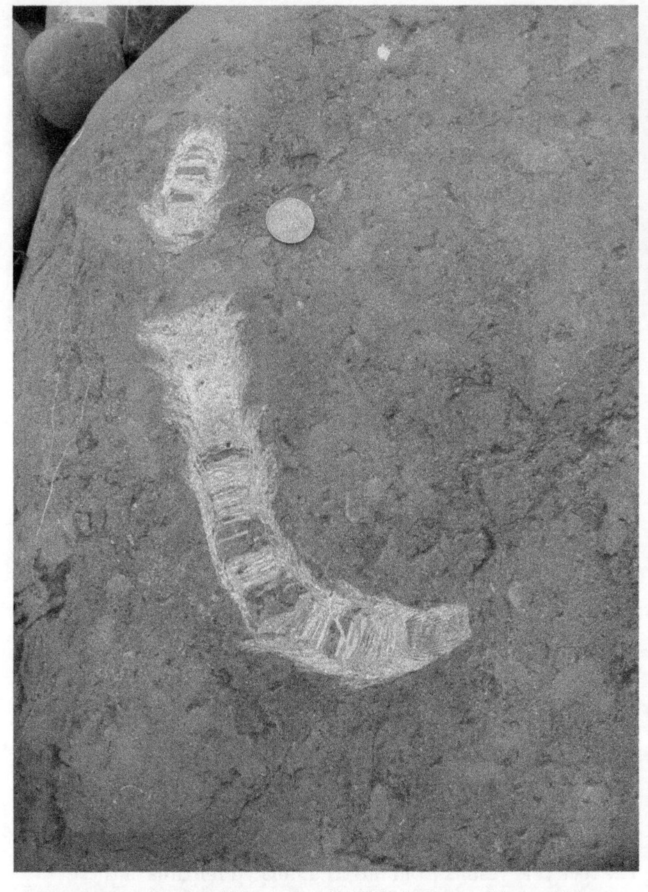

14. Limestone (Carboniferous; South Wales). Most of the rock is a carbonate mudstone, which has been thoroughly destructured on the sea floor by burrowing. It includes two prominent fossilized corals.

semi-arid, a sunny day with a drying wind will induce 'whitings' in the surface waters—spontaneous precipitation of tiny calcium carbonate crystals that then settle to the floor as carbonate mud. Many limestones have such chemically formed mud as a large component. A distinctive variation on this theme is where such precipitation takes place in shallow, current-swept oceans, and calcium carbonate precipitates in concentric layers around small mobile particles, a little like ice freezes in layers around a nucleus to form a hailstone. These are *ooids* (sometimes called *ooliths*) and they form the 'sand' grains that make up the beautiful white Bahamian beaches.

The islands of the Bahamas are only the tip of an enormous mass of limestone that extends far below sea level—100 km across and 5 km thick, it has been growing since Jurassic times, starting near the sea surface when the Atlantic Ocean was young. The underlying newly formed ocean crust was then warm and buoyant, but as it became older, colder, and denser, it subsided. To keep near the warm sunlit surface, the Bahamas had to keep building up successive layers of limestone over millions of years to attain its present gargantuan thickness. There are other such *carbonate platforms*, while atolls—as Charles Darwin was the first to realize—have formed in much the same way, the corals and other organisms building ever higher as their original foundation, an island volcano, subsided to eventually disappear beneath the waves.

Limestones are a class of rock quite different from sandstones and mudstones. Their components typically form 'in place' by biological and chemical processes, rather than arriving as far-travelled eroded particles as do silica-based pebbles, sands, and muds. And, because calcium carbonate is relatively water-soluble, it commonly changes by recrystallization quickly and profoundly soon after being deposited on the sea floor. This can make limestones rather alien and baffling rocks to geologists familiar with siliceous sedimentary rocks. One person who changed the way limestones are described and interpreted was the US geologist Robert Folk, who devised a simple yet surprisingly effective three-part classification in which all limestones could be described in terms of their contents of carbonate mud (*micrite*), carbonate cement (*sparite*), and larger components he termed *allochems* (fossils, ooids, and other particles). In his retirement, Folk studied minute fossilized particles in limestones he termed 'nannobacteria' and so found himself embroiled (to his delight, it seems) in the controversy surrounding the alleged tiny fossil bacteria in the Martian meteorite ALH84001 (see Chapter 7), a claim originally announced on the White House lawn by President Clinton. Folk's work was developed by the British scientist Robin Bathurst, that 'finest English gentleman and geologic scholar' as a colleague described him, who, at the start of his career, faced with limestones that he found 'largely incomprehensible', resolved to 'take a year off to unravel their complexities'. He spent the rest of his life pursuing this object, and transformed the field, working out such aspects as the way underground cavities can fill with crystalline calcite, or how microscopic organisms can transform calcitic fossils to a mud-like texture. He had such a close focus on geology that on one occasion, having just watched the western *Butch Cassidy and the Sundance Kid*, he could describe in detail the rocks filmed in the landscape but needed to be reminded of the plot.

On a grander scale, limestones form another of the Earth's great carbon stores—and hence act as a primary climate regulator. Comparison with our planetary neighbour, Venus, shows how. The amount of carbon that is stored in Earth's limestones, organic-rich shales, and coal is roughly equivalent to the amount of carbon in the carbon dioxide of Venus's crushingly thick atmosphere, which produces a greenhouse effect that (despite the weak light at the surface, due to thick reflective clouds of sulphuric acid droplets) keeps that planet's surface at a torrid 400°C.

It is hence a little ironic that, by burning our way through one of our great carbon stores, of oil and gas and coal, we are endangering the function of another—that of the great carbonate platforms. For with the more acid ocean waters that result from higher concentrations of carbon dioxide, it becomes progressively

15. Turbidite sandstone beds set within darker mudstone: Silurian of Aberystwyth, Wales.

more difficult to precipitate calcium carbonate. 'Deaths' of carbonate platforms and atolls have happened in the past, often associated with ancient global warming episodes. We currently represent the latest threat to the production of this form of planet-regulating rock.

Fast and slow strata

Sediment progressively travels downhill, from high ground to low, and then from land to sea, where it may form deltas or be reshaped by waves to form beaches. It can be swept by tides and storms along shallow continental shelf seas, and then much of it travels downwards again, once more firmly in the grip of gravity, to reach far along deep ocean floors. The mechanism producing this drop into the depths occurs as some shock (an earthquake, say, or powerful storm) destabilizes sediment high up on some marine slope. This means that it then pours downslope as a dense turbulent mass, before eventually coming to rest as a new sediment layer, called a *turbidite*, on the sea floor.

Such events can involve extraordinary amounts of sediment. Let us consider 'ordinary' turbidite layers of modest thickness—say 10 cm thick—which settled in a flurry of mud and sand over the sea floor in a few hours or less as sediment pours in from shallower waters (Figure 15). A modest example may have a run-out distance of 100 km and an average width of 20 km. That would give an approximate volume of 200 million cubic metres and a weight of some 500 million tons. Turbidite units can reach much greater sizes. A candidate for the world's largest turbidite layer—so far discovered, at least—is the Herodotus Basin Megaturbidite, which was formed as the sea level fell about 27,000 years ago, at the start of the Earth's most recent glacial phase. The estimated 400 km^3 of this single stratum cover some 40,000 km^2 of the Mediterranean seafloor to a depth of up to 20 metres.

In some places many successive turbidites have accumulated very rapidly on the sea floor. The Southern Uplands of Scotland are mostly made up of thick beds of greywacke—a kind of coarse muddy sandstone. These originated as sediments washed off the growing Laurentian mountains straight into an ocean trench some 450 million years ago. In this trench, strata up to 5 km thick could pile up in the order of a million years or so.

In this kind of sedimentation, mountains are reduced to rubble and flow into the ocean depths. But in other parts of that ancient sea floor, sediment accumulates continuously but *almost* imperceptibly for quite unimaginable stretches of time. Geologists call these condensed sequences. So—as thousands of metres of muddy sand poured into the Southern Uplands ocean trench, something quite different was taking place on the ocean floor beyond, barely a few hundred kilometres away. In those still depths there accumulated, perhaps a thousand times more slowly, a mixture of fine mud and organic detritus that went on to become black shales, often packed full of the fossilized remains of planktonic animals.

Here, each time span of a million years can be represented by the thickness of just 1 or 2 metres. On average, 1 mm of preserved stratum represents 1,000 years. If we allow that the original fine muds were originally 80 per cent water, this still represents just half a millimetre of sediment accumulating per century. It can seem to be a vision of near-eternity—yet these condensed successions, because they are often near-continuous, can include invaluable information on the chemistry and biology of those long-vanished oceans, and that in turn can provide clues as to the course of global climate history.

Perhaps the most slowly accumulating strata of all do not form on a seafloor, but are black metallic-looking layers that form on rocks in the arid regions. 'Desert varnish', it is called, and it is a phenomenon that puzzled Humboldt and Darwin, among many others. Slice these layers, usually less than 1 mm thick, for examination by microscope, and exceedingly thin laminae appear that are less than 1 micron (a thousandth of a millimetre) thick.

They seem to form by alternate wetting and drying. If there is a dew, a smear of silicic acid gel can form on the surface of a rock, and as this dries it can catch, like flypaper, a few tiny flakes of iron oxides (or sometimes entomb the occasional airborne microbe). Dating examples of desert varnish has shown that rates of growth ranged from 40 microns per 1,000 years to less than 1 micron per 1,000 years—or, to put it another way, less than 1 mm per million years. If one could grow a varnish layer at this rate for the whole history of the Earth, it would be a mere 4.5 metres thick. Had the varnish been growing since the beginning of the universe, it would be just 13 metres thick by now.

Rock salt

Abraham Gottlob Werner (see Chapter 2) thought that granites precipitated as crystals onto ancient sea floors. He turned out to be wrong about that—however, when seas dry up, strata of mineral salts can be produced.

With modest amounts of evaporation, carbonate minerals, notably dolomite, can form. As the brine becomes stronger, gypsum (hydrated calcium sulphate) crystallizes out, and then 'rock salt' (halite: sodium chloride, as in table salt). The most concentrated brines eventually yield chlorides of potassium and magnesium.

Some of the world's most spectacular strata of such *evaporite* minerals can exceed 2 km in thickness, as in the ~6-million-year-old salt layers that formed when the Mediterranean Sea dried up and became a hellish toxic desert. This is far more than can be derived from a single isolated water body, so such salt layers needed multiple refilling and then drying out of this Mediterranean Basin. Other such gargantuan salt deposits formed at various times and places in the world—each one in effect the result of a tectonic accident that produced just the right combination of inflow and evaporation, and their sum effect was to help regulate, over geological time, the salt content of the world's oceans.

Rocks and ice

Frozen water is a mineral, and this mineral can make up a rock, both on Earth and, very commonly, on distant planets, moons, and comets (Chapter 7). On Earth today, there are large deposits of ice strata on the cold polar regions of Antarctica and Greenland, with smaller amounts in mountain glaciers in the Rockies, Alps, and elsewhere. These ice strata, the compressed remains of annual snowfalls, have simply piled up, one above the other, over time; on Antarctica, they reach almost 5 km in thickness and at their base are about a million years old. The ice includes trapped bubbles of fossil air and particles of wind-blown dust, some of which is of volcanic origin; this adds up to a marvellous archive of information which can be interrogated to give an eloquent picture of climate change.

The ice cannot pile up for ever, however: as the pressure builds up it begins to behave plastically and to slowly flow downslope, eventually melting or, on reaching the sea, breaking off as icebergs. As the ice mass moves, it scrapes away at the underlying rock and soil, shearing these together to form a mixed deposit of mud, sand, pebbles, and characteristic striated (ice-scratched) cobbles and boulders that, as discussed earlier, used to be known as a boulder clay but is now more formally termed a *glacial till*.

Glacial tills, if found in the ancient rock record (where, hardened, they are referred to as tillites), are a sure clue to the former presence of ice. Their presence or absence through Earth's geological history betrays large-scale climate change including times of severe glaciation (such as the worldwide 'Snowball Earth' glaciations of the late Proterozoic) and the rather less severe, if still impressive, 'Ice Age' of the late Cenozoic Era (within which we still live).

Other intervals of geological time were essentially ice-free, such as the Mesozoic Era when dinosaurs lived, and in these times there was little or no ice, even at the poles. One consequence of the geological changes that humans are making to the planet (see Chapter 8) is to threaten, through global warming, to reduce the amount of ice on Earth—a change with large potential consequences for humanity.

Sedimentary rocks and time

Sedimentary rocks typically form layers (strata) that are successively laid down one on top of another. This pattern forms the basis of a 'Law of Superposition', where the oldest strata are at the bottom and the youngest at the top. This does not *always* hold true—wet, pressurized sediment underground can sometimes be squeezed upwards as sheets that intrude younger strata, and the youngest layers of a stalactite are at the growing tip (i.e. the bottom) of the structure. Rock strata can be later turned upside down by tectonic forces too. Nevertheless, the pattern is so prevalent that it forms a key guideline in reconstructing the history of the Earth.

Strata are packed with clues to the environmental conditions and age of the world in which they formed: the chemistry of the ocean basins, the climate, and, perhaps most importantly, the types of organisms then living, their remains preserved as fossils within the strata (see Chapter 6). This kind of evidence showed the early scientists that the world of the past used to be very different to the world today. For instance, the Comte de Buffon (1707–88) realized that the fossils in the rocks around his country estate in France could not be matched with any living organisms then known, and so he deduced that they represented forms of life of ancient epochs that are now extinct.

These early ideas were later elaborated and refined by scientists such as Baron Cuvier (1769–1832), William Smith (1769–1839), and Charles Lyell (1797–1875), as they developed an ever more detailed history of the Earth from the evidence in the rock strata. This kind of study is now the basis of the discipline called *stratigraphy*. This history is now tabulated in the Geological Time Scale, comprising the eons, eras, periods, epochs, and ages of the Earth (see Figure 1). Given that the Earth's history is so long and complex, this time scale is still being developed and refined by stratigraphers today. Indeed, humans are themselves now changing the course of geological history itself (see Chapter 8), and this might ultimately result in the Geological Time Scale being modified to take into account our own impact on the planet.

Making rock from sediment

If sedimentary layers have not been buried too deeply, they can remain as soft muds or loose sands for millions of years—sometimes even for hundreds of millions of years. However, most buried sedimentary layers, sooner or later, harden and turn into rock, under the combined effects of increasing heat and pressure (as they become buried ever deeper under subsequent layers of sediment) and of changes in chemical environment.

Part of the process is simply compaction as water is squeezed out and the particles are pressed more tightly together. There are also various solution and precipitation processes. For instance, where quartz grains are tightly pressed together, a process called *pressure solution* takes place, as silica dissolves away from the high-pressure contact points of the grains and diffuses to precipitate in adjacent lower pressure areas, cementing the particles together. Another common natural cement is calcium carbonate, as buried shells dissolve underground and the material subsequently precipitates around grains.

In mudrocks, originally highly porous on the sea floor, the degree of compaction through the weight of overlying strata can reach 90 per cent in some cases. These sediments too can be chemically cemented by silica, calcium carbonate, or other minerals, often just in small patches as concretions (see Figure 13) that typically form in specific buried micro-environments, such as around buried fossils. Another process is the progressive transformation of the delicate clay minerals into larger, simpler forms, a process that can involve the loss of chemically bound water. As this happens, buried organic material, already degraded by the action of microbes of the 'deep buried biosphere', begins to transform under heat and pressure. At depths of a few kilometres, where the temperatures reach and then exceed 100°C, the complex organic molecules are broken down into the simpler ones that make up oil and gas. These somehow force themselves out of the *impermeable* host mudrock and then migrate upwards, either until they reach the surface as a hydrocarbon seep, or until they are trapped in some porous and permeable rock, such as a sandstone with open pores between the grains, or a fractured and cavity-rich limestone. There the oil waits until humans find and extract it.

As rocks become buried ever deeper, they become progressively changed. At some stage, they begin to change their character and depart from the condition of sedimentary strata. At this point, usually beginning several kilometres below the surface, buried igneous rocks begin to transform too. The process of metamorphism has started, and may progress until those original strata become quite unrecognizable. This, a further next stage in the rock cycle, is the process we shall explore next.

Chapter 4
Rocks transformed

Mountain belts do not just raise masses of rock high in the air to form jagged peaks traversed by deep gorges. They also represent substantial thickenings of the crust. The rocks at the bottom of such thickened crustal masses may lie tens of kilometres below ground for many millions of years, and then slowly rise to the surface again as erosion and weathering wears away at the rocks overlying them.

Part of the thickness increase is original—that is, the rocks have not only been crumpled into a thicker mass, they have formed thicker successions of strata even before the crumpling began. The early geologists, puzzled by this phenomenon, described it in terms of geosynclines—long, thin, rapidly subsiding marine basins that somehow then were compressed to form mountain ranges. We now know that those 'geosynclines' were ocean trenches and associated sedimentary basins linked with subduction zones, where tectonic plates converged. The basins collected sediment from the rising mountains, and then were compressed into mountains themselves, as the tectonic plates maintained their inexorable movement. The rocks caught in this tectonic vice are subject to enormous pressures and increased temperatures, and this is a zone in which they are transformed. This is the realm of regional metamorphism.

The outer zones

In the outer zones of a mountain belt, the rocks of the crumple zone may reach depths of 5 km or more, and temperatures may rise to some 200°C and more. This temperature rise is about 25°C with every kilometre of descent to lower levels of continental crust. It reflects the inner heat of the Earth, mostly generated by the breakdown of radioactive elements in the Earth, which escapes only slowly because rock is a good insulator (the mass of rock of the Earth is such a good insulator that some heat still remains from the violent collisions associated with its formation, four and a half billion years ago). As the temperature rises, the complex, delicate mineralogy of the mudrocks expresses the changes most noticeably. At these temperatures and pressures, the clay minerals carry on with the transformations that they had started in sedimentary burial: they are progressively converted into crystals of the mica family that are, at first, tiny (though not as tiny as the original clay minerals) and then grow larger.

The recrystallization may simply mimic the original shape and alignment of the former clay minerals. More commonly, though, the new tectonic stress field associated with squeezing and folding

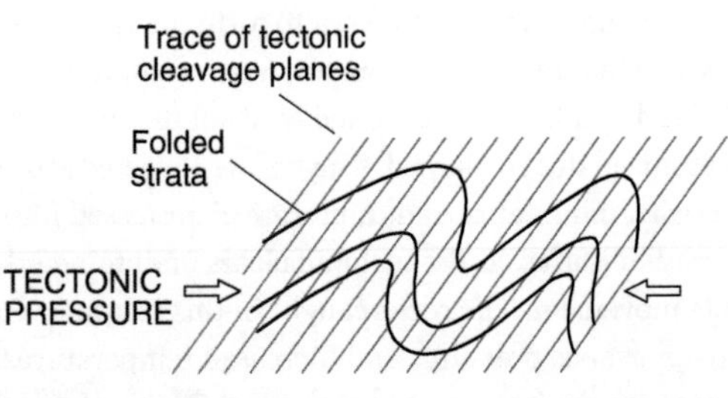

16. Diagram showing how slaty cleavage develops in compressed and folded mudrock strata.

the rocks controls the new crystal growth, as the entire crust is shortened. The tiny new micas grow perpendicular to the stress field and so parallel to the hinge planes of the developing rock folds (Figure 16). This reorientation of the rock fabric means that the rock no longer splits along the plane of the stratification, as it had tended to previously, but along *tectonic cleavage planes* formed by the new mica crystals.

The mudrock, so transformed or metamorphosed, has become a slate (Figure 17). When the pressure has been evenly exerted on a homogenous mudrock, the new fabric can be beautifully smooth—smooth enough to cleave into slabs only a quarter of an inch thick (by skilled hands, mind) for roofing slates, or for a high-quality, albeit heavy, billiard table. Traces of the original sedimentary layering can often be seen to cross the surface at an angle, though if this layering involves a change in grain size to a silt- or sandstone, then the splitting surfaces are uneven and the slate is no longer of commercial quality.

The internal texture of a slate can be beautiful, too. In some rocks, the mud may originally have had within it flakes of eroded mica that had drifted onto the sea floor with the mud. These micas can be stretched by the tectonic forces and opened out like the pages of a book, with new mica growing in the spaces so created. This occurs in such a way that the mica flake becomes a newly grown, barrel-shaped, metamorphic mineral. Fossils within the mud can be altered too. At such a low grade of metamorphism they are not yet obliterated, but—especially if they are made of a hard brittle mineral such as pyrite—they can be stretched into a series of broken segments; or, if squashed, those segments can be pushed one on top of another. It makes for a fascinating jigsaw puzzle for the palaeontologist to reconstruct.

As temperatures and pressures rise still further, there is an increase in metamorphic grade (Figure 18). The new micas

17. Slate showing the near-vertical tectonic cleavage planes, cutting across the original stratification, which is more faintly seen and which dips gently towards the hammer. (Silurian of Rhayader, Wales.)

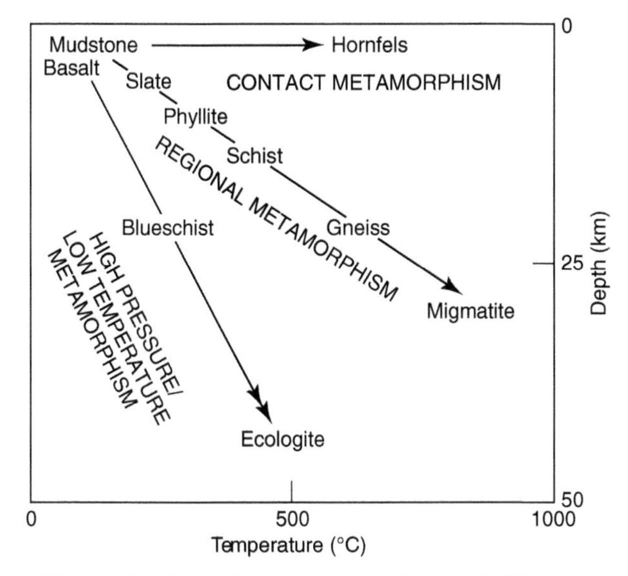

18. Diagram showing pathways of metamorphism with different trajectories of temperature and pressure.

continue to grow larger. In a slate they are still so tiny that rock surfaces have a dull appearance, but once they become a little larger they give the whole rock a lustrous sheen. The rock has now become a phyllite, at temperatures of around 400°C and depths exceeding 10 km. It is not just the micas; other minerals, like the quartz, are recrystallizing too, while any scattered small grains of pyrite in the rock refashion themselves into golden-coloured cubic crystals that can be 1 cm or more in size.

The middle regions

Ratchet up the temperature and pressure yet further, on moving closer towards the heart of a mountain belt, and the crystals of mica become big enough to see with the naked eye. The rock has

19. Microscope thin section of a garnet mica-schist. The fibrous-looking micas and the rounded dark garnets have grown in the solid state under heat and pressure generated in the roots of a mountain belt.

become a mica-schist, a metamorphic rock in which the mica-rich layers alternate with quartz-rich layers. This layering is now nothing at all to do with the original stratification, which has mostly been obliterated, together with any remaining fossils, but reflects chemical segregation in the ever hotter and more highly pressurized environment. New minerals form here, notably garnet (Figure 19), its near-spherical crystals often containing spiral-shaped trails of mineral inclusions that show how it rotated in response to the shearing forces around it. As temperatures rise, the garnet is replaced by the minerals staurolite, kyanite, and then silliminite—all aluminosilicates—and now classic indicators of successive zones of increasing temperature and pressure.

This environment is harder to feel and visualize than the surface of the planet Venus, or of a speeding comet. If a cavern could be opened up within such a realm—and if any human observers (suitably armoured) dared briefly to step inside, then they would need to use torches to see the walls, as the glow of the rock, even at these temperatures, is too far in the infra-red to be seen by human eyes. Great patience would be needed to see any of these mineral transformations, for they happen (we think), exceedingly slowly, over many millennia and even millions of years. These mineral changes take place in a solid state (for we are at temperatures still far short of melting) and within solid rock. They are mediated by tiny amounts of fluid (mostly water) that permeate—terribly slowly—along the imperceptibly slowly evolving mineral grain boundaries, in this rock that has been squeezed *almost* dry by the pressures upon it. For all practical purposes, it is an environment as alien and inhospitable to a living carbon-based organism as is the heart of a star.

Heart of a mountain belt

Deeper inside the mountain belt, at temperatures of around 600°C and at depths of more than 20 km, the mineral transformations continue. Much of the mica itself becomes unstable and breaks down, while, instead, feldspar becomes a large component of the rock. The mineralogy and crystalline texture here resembles that of an igneous rock—granite—yet is still the result of solid-state transformations. The rock is a gneiss (Figure 20), which may be distinguished from a granite by banding caused by mineral segregation and shearing. A related rock, granulite, has less banding and may closely resemble a granite. Indeed, at this stage of metamorphism it is difficult to distinguish gneisses that were originally sedimentary rocks from those that were igneous (such as granites with a texture modified by shearing processes).

The next stage, indeed, is melting, and there is a class of rocks—the migmatites—in which layers of melt alternate with layers of still largely unmelted, high-grade metamorphic rock, before the complete melting takes place that will lead to the formation of magmas, the composition of which will depend on the nature of

20. Gneiss, India.

the original rock, the so-called *protolith*. The rocks have now gone full circle, from the primary igneous rock to their surface weathering and breakdown into sedimentary detritus that will become sedimentary rocks, to the progressive burial and metamorphism of these until the formation, once more, of a magma. The cycle (Figure 21) is complete.

The particular kinds of rocks that form along this route differ depending on their composition. We have been talking mainly about 'average' continental crustal rocks, which may be very loosely represented by a mudstone on the sedimentary side and by granite on the igneous rock side. However, a body of more or less pure quartz sandstone cannot form either a slate or a schist because it lacks the mineral components that could develop into micas. Here, the quartz grains simply become ever more tightly compressed and recrystallized into a mosaic-type structure (albeit a three-dimensional one): the metamorphic rock that forms is a quartzite. Limestones, meanwhile, recrystallize into forms of marble, and in the process they progressively lose both their

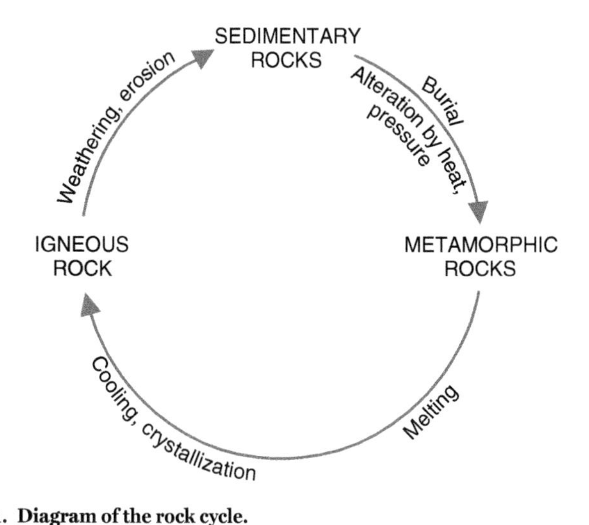

21. Diagram of the rock cycle.

original sedimentary textures and their recognizable fossil content. Metamorphism may be a great leveller of original diversity, but the primary categories defined on chemical composition are still reflected within this new family of rocks.

The cool path in metamorphism

The metamorphic progression described earlier is a classical 'regional' one of mountain belts, where crustal thickening and hence progressive burial heating is a key factor. However, it has long been known that mountain belts can also include elongated zones of metamorphic rocks that have not been heated so strongly, but that show evidence of far greater pressures—where rocks have been transported to depths of up to 100 km (or more) before being exhumed back to the surface, and often at a somewhat unseemly speed for such deep crustal or even mantle depths.

These rocks do not generally show evidence of moving towards a melt state, rather they show unmistakable signs of having experienced very high pressures, in which grew minerals such as the distinctively blue glaucophane (typical of *blueschist* rocks) and green jadeite and even, in places, tiny diamonds. The rock serpentinite is part of this family, with its chief mineral, serpentine, being a lustrous green iron- and magnesium-rich silicate characteristic of the kind of *ultramafic* rocks of the mantle that are the source rocks for basalt. Serpentinite not only has this ultramafic composition but is also hydrated, with water molecules bound into its structure, showing that it clearly formed in a very fluid-rich as well as a high-pressure environment. The dense, distinctive, red-and-green garnet-rich rock eclogite may also be found here—it represents a high-pressure, but relatively cool, transformation of basalt. So what is going on here?

This enigmatic set of observations fell into place once plate tectonics appeared as a fundamental framework in geology, and these strange low-temperature, high-pressure, metamorphic rocks are now clearly associated with the descending plate of the subduction zone. The high pressure simply reflects the depths to which such a slab can descend and still release fragments that return to the surface. In reality the slab goes much further down, but the really deep portions are, with rare exceptions (see Chapter 5), irretrievably lodged in the depths of the Earth. The relative coolness of the transformations reflects the fact that the descending slab of oceanic crust has—after its tens of millions of years of slowly inching across the expanse of ocean basin—now cooled substantially from its original high temperature as a magma at the mid-ocean ridge. And the evidence of the involvement of substantial amounts of water stems from the surface of the ocean crust itself, together with the overlying sediments, being water-soaked. This water is mostly wrung out as the slab descends, but as it makes its tortuous way back to the surface it takes part in the mineral reactions of this kind of metamorphism.

Time, pressure, and temperature

Time is the backbone of all of geology: to reconstruct the enormously long history of this planet, one needs to order the component phenomena effectively in time and space, so as not to put together chaotic and nonsensical sequences of events. Hence the need to assign an age to the different rocks that one might be studying. With igneous rocks, the 'age' is generally understood as the time at which the rock has crystallized from a magma to a solid; and for a sedimentary rock it is the time that the particles finally settled on the sea floor or river bed.

For a metamorphic rock, though, this kind of simple pigeon-holing is more difficult to achieve. The rock has in effect taken a path of increasing metamorphic grade (though with various combinations of high and low pressures and temperatures). Then, it has slowly cooled and undergone pressure release as those rocks, over millions of years, have made their way back to the surface, normally via the erosion of several kilometres of overlying strata. The resulting rocks can preserve evidence not just of the *peak metamorphism* but also of various stages along the path of *prograde metamorphism* (leading to that peak), and of the subsequent *retrograde metamorphism* (as part of the mineral assemblages may convert into forms more at equilibrium with the decreasing temperatures and pressures on their way back to the surface). Some of the mineral phases formed along this journey may be possible to date by analysis of their natural radioactivity (see Chapter 2). For instance, small amounts of metamorphic forms of the minerals monazite and xenotime may be present (these are rare earth element phosphates that may include substantial amounts of uranium when they crystallize). Hence a kind of four-dimensional path may be reconstructed for a single metamorphic rock along what is known as a *pressure-temperature-time curve*, illustrating the course of its long journey through the Earth's crust. At any stage in this process, fluids wrung from the

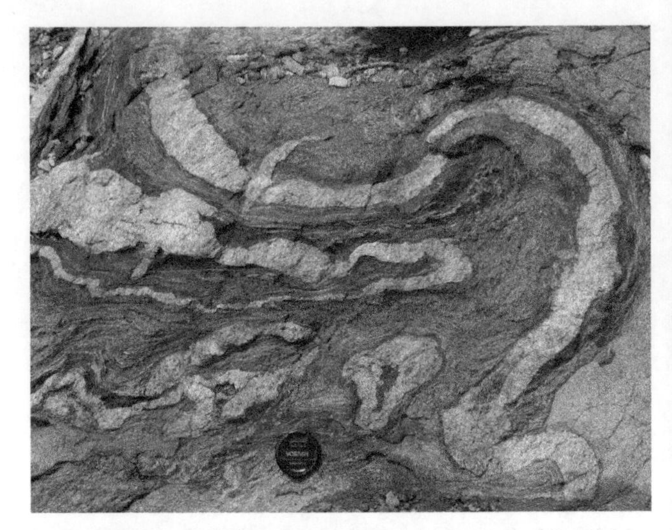

22. Crumpled and cross-cutting quartz veins reflect a complex history of rock deformation and fluid release during metamorphism. (**Cap de Creus, Spain.**)

rocks may be a source of minerals that can precipitate in fractures opened up by tectonic forces; such mineral veins, which can have their own complex histories (Figure 22), are often dominated by quartz, but they may also be enriched in metals such as copper, lead, tin—and gold.

Contact metamorphism

Most of the metamorphic rocks of the world are associated with mountain building processes, as described earlier. But there is a particular class that reflects just the action of heat, with little added pressure. This is where a magma body comes into contact with a rock mass, and the heat from the magma bakes and recrystallizes the adjacent rock.

The extent of such contact metamorphism can vary greatly. Where magma penetrates as a thin sheet into a fracture or plane of weakness between adjacent sedimentary strata, the amount of heat transfer and hence 'baking' can be slight or even imperceptible. However, at the edge of a substantial magma chamber, the region of alteration—known as a *metamorphic aureole*—can extend for tens or even hundreds of metres into the rock. The scale of contact metamorphism in such cases may reflect the extent of fluid transfer from the magma as much, or more, than it does the simple conduction of heat through dry rock.

The kind of rock formed depends to a large extent on the degree of heating and on the nature of the rock being heated. Mafic and ultramafic magmas, with their higher melting points, will heat rocks more strongly than granites—though granite bodies can be huge, and their corresponding metamorphic aureoles can be substantial.

Contact metamorphism of a mudrock or slate will first produce 'spots' in the rock—small patches of recrystallized aluminosilicate minerals. With further heating, some of those grow into the distinctive elongated crystals of chiastolite. These are so regular that they superficially resemble fossils, not least because of the dark cross seen when a crystal is broken across—a cross made of impurities forced into this central pattern in the growing crystal. The next stage is a hornfels, a thoroughly recrystallized rock, so named after its supposed resemblance to animal horn.

Hornfels has one rather unexpected quality—when suitably shaped, it can produce beautiful musical notes when struck. Indeed, it took central place in an extraordinary narrative of the English Lake District. An eccentric 18th-century inventor, Peter Crosthwaite—a fighter against Malay pirates in his youth and, later in life, the founder of a museum in the town of Keswick—built a kind of xylophone using hornfels from the local Skiddaw mountain. Half a century later, the Keswick stone-maker and musician Joseph Richardson determined to top Crosthwaite's achievement, and almost ruined his family financially by building an even bigger instrument, which would produce a larger range of musical notes. Once built, though, it was indeed a sensation. Richardson toured England for three years with his sons, playing Handel, Mozart, and dance tunes on his rock creation—though at times restraining the power of the instrument so it would not shatter concert hall windows. Queen Victoria liked the performances so much that she requested extra concerts (although reports from the time do suggest that she was not amused at its imitation of Alpine bells). The harmonious hornfels 'lithophones' may still be seen in the Keswick museum—and are to this day occasionally taken on musical tour.

Impact metamorphism

While some kinds of rock, like hornfels, can make a ringing sound when struck with a hammer, far greater impacts can occur in nature, and these can create not just sound waves (though in these cases the whole Earth may ring like a bell) but also, in an instant, their own distinctive kind of metamorphic rock.

The pretty, walled town of Nordlingen in Germany is set within an almost perfectly flat plain with a circular line of hills surrounding it like ramparts about 10 km away on all sides. These are not ancient defences constructed by the townspeople: far older than the town itself, they are the walls of a crater that formed a little over fourteen million years ago. The crater used to be thought to have been volcanic, but the rock that has been quarried there for aggregate to use in the town looks like no ordinary volcanic rock. It is a broken mixture of finely crushed local bedrock and melted, bubble-filled fragments that is in fact a product of impact: it formed when an asteroid over 1 km across slammed into the ground late in the Miocene Epoch. The event caused immense local damage, choked the local river systems with debris, and sent showers of melt (now preserved as the fossilized droplets termed *tektites*) as far away as Moravia; however, it seems not to have precipitated any major change to the global biosphere.

The key clues confirming that this shattered local rock is indeed an *impactite* lie in tell-tale minerals preserved within it. There is a high-pressure form of quartz, called coesite, and also millions of tiny diamonds scattered within it—and so also present within the walls of the buildings constructed from this rock. The amount of diamond produced by the shock waves has been estimated at 70,000 tons, the abundance of this mineral reflecting the direct hit of the asteroid on a graphite deposit.

Impactites are associated with the large identified meteorite craters on Earth, including the one, some 200 km across, that lies buried beneath the Yucatan Peninsula of Mexico, the formation of which sixty-five million years ago finished off the dinosaurs and much else of the Earth's biota. Within such large craters, the impactite is dominated by melted and deformed rock. In the case of the Yucatan crater, such deformed rocks extended right through the crust into the mantle.

The material that can spill outside such craters is just as remarkable. There is a billion-year-old rock stratum up to 20 metres thick that extends for up to 50 km from close to a picturesque sea-stack called Stac Fada in the far north-west of Scotland. It looks just like the deposit of a pyroclastic current from an enormous volcanic eruption, but it contains the tell-tale high-pressure form of quartz and other clues betraying an impact origin. It is in effect not so much a pyroclastic current as what has been called an 'impactoclastic current' of material that was driven outwards from the point of impact as a dense, debris-packed, ground-hugging flow. If one had then been in what is now Aberdeen, on the other side of Scotland, the winds experienced would (it has been estimated) have exceeded 400 km an hour.

Chapter 5
Rocks in the deep

We only *really* know the exterior of our planet, rather like the microbes that cling only to the surface of the skin of an apple, oblivious to the flesh of the apple that lies within. The Earth is a very large mass of rock, being over 6,000 km to its centre. Our direct experience of it goes to almost 4 km below the surface—in the deepest mines on Earth, the South African gold mines. The crust there is comparatively cool, which allows humans to penetrate that far, albeit aided by excellent cooling and ventilation systems, and skilled engineering to counter the ever-present danger of rockbursts and cave-ins associated with those crushing pressures. Indirectly we can penetrate further by drilling, and oil boreholes commonly extend to 5 km and more. However, the record depth attained is still the 12 km reached in 1989 in the Kola Super-Deep Borehole in Russia, which, terminating in rocks at 180°C, only penetrated to about one-third the thickness of the local continental crust.

To go deeper we have to cheat, and exploit the rock fragments brought from deeper levels by tectonic or volcanic processes. Alternatively, we can probe the Earth by means of analysing the patterns of change in its gravitational and magnetic fields; or by detecting seismic waves that have travelled through the Earth. Or, we try to re-create the conditions of the deep Earth in the laboratory (astonishingly, with some success).

It is now quite clear that the depths of the Earth are nothing—alas!—like the vistas dreamed up by Jules Verne in his *Journey to the Centre of the Earth*, in which tortuous caves lead to underground lands with dinosaurs and giant humans, and to seas where ichthyosaurs do battle with plesiosaurs. The depths of the Earth are the province of rock that becomes progressively more different from that of the oxidized, weathered, crumbled material at the surface. It is where the real stuff of a planet resides.

The Kola borehole itself was expected to pass from granite to that denser rock, basalt. However, that never happened, and the granite persisted to the point where drilling was abandoned. To get to the denser rocks of the interior—indeed to the ultrafic rocks from which crustal basalts are derived—one needs to go through the crust and penetrate the Moho boundary (see Chapter 2) that separates it from the mantle. The Moho typically appears as a sharp increase in rock density when detected by analysis of seismic waves. This has never yet been drilled through in present-day crust, not even in the oceans where the Moho lies at a depth of only between 5 and 10 km.

However, nature has already done the job for us, as tectonic displacements have, here and there, dislocated large slabs of lithosphere (most typically as the detached oceanic slivers termed *ophiolites*) and pushed them up into higher crustal levels—in places exposing ancient examples of the Moho boundary in the process. This shows that the sharp boundary seen by geophysics is more complex and gradational in the rock—but no less real for that. The 'normal' basalts and their coarser equivalents, gabbros, of the lowermost crust pass downwards into the ultramafic rocks of the upper mantle. These have even less silica and more iron and magnesium than do mafic rocks, which accounts for the density increase across the Moho. The main rock type here is peridotite, which is made up of the mineral olivine together with varying amounts of pyroxenes. There may be a little feldspar, but that mineral, at these levels, no longer has anything like the dominance that it shows in the crust. A little deeper in the mantle, it disappears altogether, its relatively open structure collapsing down and being replaced by denser mineral forms such as spinel.

These are the rocks that make up the upper mantle, and which act as a source, by partial melting, for the voluminous basalt magmas that make up the ocean crust (see Chapter 2), or that can at times flood across continents. The Moho can be seen as this transition from mafic to ultramafic rocks—on Ślęża mountain, in south-west Poland, for instance; and in the Hajar mountains of Oman—and in such places it takes only a few minutes to cross it on foot.

One does not have to go to such special—now almost iconic—places to see mantle rock. If one is happy to see it as fist-sized lumps, then these can be encountered in basalt rocks generally, particularly in oceanic settings. They are *xenoliths* ('foreign rocks'): fragments that have been detached from the margins of mantle-hosted magma conduits and carried to the surface during eruptions. Xenoliths can be striking in appearance, as for instance when the bottle-green olivine that makes up most of a mantle peridotite contrasts strongly with the dark tones of a basalt (Figure 23). Close examination of such messengers from the deep can give important clues to the state and behaviour of the mantle beneath a volcanic system. For instance, the mantle below the large volcanic island of Tenerife (the third biggest volcano in the world) was shown by xenoliths to be extensively altered by calcium-bearing fluids that had travelled up from greater mantle depths, while the mantle itself had undergone a greater-than-usual degree of partial melting as the North Atlantic Ocean in that region was splitting open.

The diamond factory

Natural sampling of this kind can range deeper than the Moho, while for the deepest-derived rocks of all one seeks a special kind

23. Peridotite (green) mantle xenoliths within (dark) basalt from Arizona, USA.

of volcano. It is so special that there are no historical records of eruptions (though they must have been spectacular, in the far-off days when they did take place). These are kimberlites, the distinctive fragmental rocks that provide the 'blue clay' of diamond pipes. The diamonds and other mineral rock fragments that they carry to the surface were derived from depths as great as 200 km—well into the mantle.

How did these rocks and minerals arrive at the surface while still preserving their deep-mantle character? Very quickly, it seems. A propulsive power is needed, and the chief suspect here is carbon dioxide, aided and abetted by that usual volatile substance, water. Some special kind of source material in the mantle seems to be required, and one suspect is subducted ocean crust with its sediment cover carrying carbon-rich material deep below the Earth's surface, to be converted into the characteristic kimberlite mineral—diamond—and also the carbon dioxide propellant.

The kind of speeds of ascent of this mantle material have been estimated at some hundreds of metres a second, suggesting that kimberlite eruptions must have been spectacular. Some of the more fanciful interpretations have the outrushing gas-charged magma propelling a column of rock from the carrot-shaped vent so far and fast that its landfall (termed a 'Verneshot' after another of Jules Verne's classic stories: *Voyage to the Moon*) mimics the effect of a catastrophic meteorite impact. A little more soberly, the mantle-derived carbon dioxide released from kimberlite eruptions—which have occurred very sporadically in Earth history—has been linked (rather speculatively, it must be said) with ancient episodes of global warming such as the brief, brutal 'hyperthermal' at the end of the Paleocene Epoch: the Paleocene-Eocene Thermal Maximum or PETM that occurred about fifty-five million years ago.

Another mineral fragment catapulted up to the surface by a kimberlite eruption has been a tiny fragment of ringwoodite (a high-pressure form of olivine). It was highly significant in a different way. Smuggled within a battered diamond on an enormous ascent, even for kimberlites, of more than 500 km, this mineral tells a story of the water stored in Earth's deep mantle rock. When analysed, it was found to be composed of over 1 per cent water. This suggested that the mantle in this deep *transition zone* between the upper and lower mantle holds over an ocean's worth of water, with implications for rock properties in this deep zone: the more dissolved water there is, the less rigid the rock, allowing more deformable and plastic behaviour than would be possible within a drier mantle. It is a clue to a deep Earth that seems all the more dynamic the more closely it is studied.

The deep mantle

For a long time there has seemed to be no direct way for rock to arrive to the surface from levels deeper than the transition zone that separates the upper mantle from the lower mantle. The rocks of the lower mantle are even denser than those above the transition zone, as inferred from patterns of seismic waves and also from laboratory experiments. To make such rocks in the laboratory needs the jaws of a diamond anvil to focus the kind of enormous pressure required onto a pinch of normal surface mineral in order to mimic the conditions experienced in deep Earth. Then, shining X-rays through it allows its molecular structure to be worked out. The high-pressure minerals thus formed—such as ferropericlase and perovskite—were thought to be relatively homogeneously distributed in a generally uniform lower mantle.

However, exceedingly rare diamonds have been found to contain lower mantle minerals, thus extending the roots of (occasional) volcanic eruptions to almost unimaginable depths in the Earth. These minerals, sparsely found in kimberlites worldwide, suggest a more complex structure to the lower mantle than had previously been thought. Free silica can also be separated out from the rocks found in these depths, in the form of the high-pressure mineral stishovite.

Moving rocks

At first approximation, the mantle is made of solid rock and is not—like in many old storybook pictures of the Earth—a seething mass of magma that the fragile crust threatens to founder into. This solidity is maintained despite temperatures that, towards the base of the mantle, are of the order of 3,000°C—temperatures that would *very* easily melt rock at the surface. It is the immense pressures deep in the Earth, increasing more or less in step with temperature, that keep the mantle rock in solid form. In more detail, the solid rock of the mantle may include greater or lesser (but usually lesser) amounts of melted material, which locally can gather to produce magma chambers (or, if turbo-charged with dissolved carbon dioxide, can speed from deep mantle levels to the surface as kimberlites, as we have seen).

Nevertheless, the mantle rock is not solid in the sense that we might imagine at the surface: it is mobile, and much of it is slowly moving plastically, taking long journeys that, over many millions of years, may encompass the entire thickness of the mantle (the kinds of speeds estimated are comparable to those at which tectonic plates move, of a few centimetres a year). These are the movements that drive plate tectonics and that, in turn, are driven by the variation in temperature (and therefore density) from the contact region with the hot core, to the cooler regions of the upper mantle.

These movements include large, slow, convection currents, over which there has been much debate: do they form separate loops in upper and lower mantle or do they pass through the full thickness of mantle (which now seems to be more likely)? The convection currents come near the Earth's surface and separate at or around mid-ocean ridges, the upwelling mantle material partially melting to produce the basaltic lavas of the ocean crust. To compensate, cooler, denser mantle rock is slowly pouring back towards the core regions and, with it, along subduction zones, come descending slabs of cold lithospheric plate. These maintain their integrity as distinct units for quite some time, perhaps—as we shall see later—even as far as the core–mantle boundary, but their fate is to eventually mix and reintegrate with mantle rock.

There are two regions where the mantle rock moves more freely. One is near, but not quite at, the very top of the mantle. This is the weaker and more plastic zone, somewhere between 100 and 200 km below the Earth's surface, called the *asthenosphere*. It contrasts markedly with the rigid lithosphere that makes up the moving tectonic plates above, and indeed it allows the movement of these plates. What makes this particular band of mantle rock so weak and mobile? Two inter-related factors are usually cited. The first is the elevated levels of water dissolved in the mantle rock at these levels; this would make the rock yet more plastic and easily deformable, and also lower its melting point. The second factor is being in a zone where the increase in heat (tending to melt the rock) is outpacing the increase in pressure (tending to keep it solid). This would provide elevated levels of melt in the rock and make it more deformable. It is a fine balance—too much melt and it would separate and migrate up towards the surface, leaving a residue that is more rigid than the original material.

The other area of more mobile mantle rock is in rising plumes—more or less cylindrical rising columns of hotter-than-average mantle rock, perhaps 100 km or so in diameter, that rise from the depths of the Earth to 'fire' on the base of the crust like a blowtorch, generating extra magma and therefore large volcanoes. The classic example, as we have seen, is the Hawai'ian line of volcanoes. Over several million years this has successively flared up and then died as the ocean crust it is on passed over the impinging plume top, culminating in the enormous, and indeed the world's biggest, present-day (and still active) volcano that makes up much of Hawai'i itself.

There has been much discussion over whether mantle plumes are real or not. One school of thought has held that they did not exist, and exactly the same effect could be caused by a hotter-than-usual patch of the upper mantle acting on the base of the plate that was moving above it. It is impossible to journey down into the mantle to find out for sure, after all. However, our ability to read the faint, geophysical signals that emerge from the solid Earth, for example through the interpretation of earthquake wave patterns and the variations in the Earth's gravitational and magnetic fields, has increased to the point of capturing hints of what are interpreted as very deep-lying, near-vertical, pipe-like structures of subtly different physical rock properties. These, then, do seem to be imaged traces of mantle plumes.

How deep do they go? And what is the structure of the broader scale mantle convection currents? Some plumes, at least, are thought to rise from hotter parts of the core–mantle boundary and travel upwards (albeit being pushed this way and that by the slow-moving 'weather' of the mantle convection currents) to finally impinge on the base of the lithosphere and spread out, as the tops of >1,000-km-high rock fountains. There they generate volcanism and may physically lift a large area of crust by a kilometre or more. There is, beneath the North Sea, a fifty-five-million-year-old buried landscape with deep canyons thought to have been carved during just such a plume-generated uplift event, when part of the North Atlantic Ocean began to open.

Mantle currents, as they pass through the mantle transition zone, will undergo a mineral phase change, from the olivine-pyroxene-garnet assemblage of the upper mantle to the perovskite composition of the lower mantle on descending, or vice versa on ascending.

The mantle currents form an enormous, deep-running, planetary machine (Figure 24), and its effects on the surface may be seen as the familiar, yet ever-renewed geography of continents and oceans, mountain ranges and plains. As the Earth overall is cooling, still losing stored heat that was generated from its cataclysmic assembly 4.5 billion years ago, and as its internal radioactive heat declines, its motor of mobile rock is thought to be slowing down.

However, there are tantalizing signs that it may rather have been speeding up over the past billion or so years, as seen by a slight increase in the rate of formation of mountain belts detected in some studies. How so? The agent here may be extra water carried down by subducting plates, very slowly mixing into the mantle rock and reducing its viscosity. There is a balance here between the water carried down into the Earth and that erupting out as steam in volcanoes. It is a difficult balance to measure scientifically, but it does seem that, long-term, there is progressively less water in the oceans and more in the mantle, which may help the Earth's plate tectonic mechanism to operate more freely.

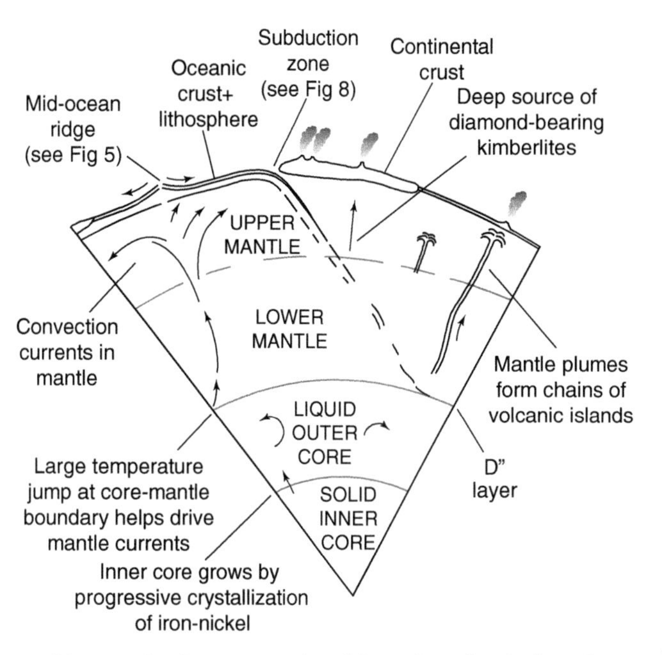

24. Diagram showing cross-section of the main rock units from the crust to the centre of the Earth.

The very deepest mantle rock

For some time, it has been realized that something odd occurs at the core–mantle boundary, where earthquake waves would change both speed and direction markedly. This mysterious layer, estimated at just a couple of hundred kilometres thick, was given the label D″ ('D-double-prime') and it has attracted great attention.

Its physical nature was established in the early part of this millennium, as the diamond anvils in the mineralogy laboratories became capable of re-creating the depths and pressures of the core–mantle boundary. The perovskite of the upper mantle changed again to the more closely packed molecular structure of a mineral that was simply named 'post-perovskite'. It is the distinctive layer with this mineral, hence, that modulates the structure of the core–mantle boundary.

Yet, there is much more going on here than just a pressure-dependent phase transition. The core–mantle boundary has been called the most important boundary on this planet because of a marked temperature jump from something like 3,000°C in the lower mantle to around 4,000°C in the outer core. This is accompanied by a sharp change in physical properties from mostly solid rock mantle to dense liquid iron core, of a viscosity that may be not very much greater than that of water. The boundary region is also complex because it marks the resting place (or 'graveyard', as it has been called) of subducted slabs of ocean crust that have traversed the full thickness of the mantle. Not quite the final resting place: for the heat of the core eventually warms these cold slabs to the point where part of its substance begins to ascend again as the initial stages of convection currents and mantle plumes.

The slabs, in descending, also seem to deform the D″ layer, so that the post-perovskite crystals become aligned. This alignment has been deduced from the behaviour of the seismic waves passing through this layer, as the waves have different speeds that are thought to reflect the different paths they take through these aligned crystal masses. The rock structure of this region is made yet more diverse by piled masses of dense, probably iron-rich material forming 'basal mélanges' in this region (these too are inferred from the complex seismic patterns). This dense material perhaps includes iron-rich segregations from the initial cooling of the Earth; more recent segregations from tectonic slabs in the 'graveyard'; and remnants of other processes through the intervening several billion years of Earth history. The rocks of the interior may be only accessible to us via distant geophysical echoes and laboratory experiments, yet even these barely adequate means hint at a richness of rock structure perhaps rivalling that on our planet's surface.

The core

The core lies a little less than 3,000 km below our feet. That is, less than a hundredth of the distance that separates us from the Moon. Yet we know the Moon's surface intimately now, and humans have walked—even played golf—on its surface.

Could we ever have anything like this familiarity with the deep and dense heart of our planet? Well, it has been suggested that a voyage like that undertaken in that Hollywood blockbuster, *The Core*, might be possible—albeit not with human passengers, no matter how heroic. The mechanism would involve making a *very* large amount of molten iron and then letting that sink under its weight through both crust and mantle, its weight creating a fracture that would then rapidly (at about 5 metres a second) propagate downwards, closing up again once the mass of iron has poured through. Place a neutrally buoyant probe in the iron, and it would be carried downwards to the core, signalling the data gathered by its sensors via high-frequency seismic waves gathered by a detector on the surface. Far-fetched? Of course—but perhaps, just perhaps, it is technically possible—given an investment comparable to that for the space programme (which the author of this proposal did admit was unlikely).

In the absence of such an enterprise, one has to do one's best to reconstruct the core based on considerations of planetary mass and seismic waves, together with information from nickel-iron meteorites (representing primitive 'core' material of shattered planetary bodies).

The outer core will not transmit certain types of seismic waves, which indicates that it is molten. Currents in this low-viscosity molten metal are thought to generate the Earth's magnetic field, which provides a protective magnetosphere to shield the atmosphere and oceans from the worst effects of the solar wind. (The Earth's magnetic field is also imprinted onto rocks at the Earth's surface, in effect as tiny, frozen-in, natural compass needles, allowing geologists to use them to track the motion of tectonic plates across the Earth). It is very hot, this molten metal, given the jump in temperature on crossing into the core from the mantle, which is of the order of 1,000°C. It is contact with this molten inferno that drives the rock currents of the mantle.

Even farther into the interior, at the heart of the Earth, this metal magma becomes rock once more, albeit a rock that is mostly crystalline iron and nickel. However, it was not always so. The core used to be liquid throughout and then, some time ago, it began to crystallize into iron-nickel rock. Quite when this happened has been widely debated, with estimates ranging from over three billion years ago to about half a billion years ago. The inner core has now grown to something like 2,400 km across. Even allowing for the huge spans of geological time involved, this implies estimated rates of solidification that are impressive in real time—of some thousands of tons of molten metal crystallizing into solid form *per second*.

The first person to 'see' the Earth's inner core was a remarkable woman, the Danish seismologist Inge Lehmann. Brought up in Copenhagen in the early 20th century to believe, as a schoolchild, that there was no difference in intellectual ability between the sexes, she noted having some 'disappointments later in life' when she saw that this view was not always shared. This might help explain her reputation for 'not always [being] very diplomatic', as her nephew later said of her. Diplomatic or not, she was a fine scientist and mathematician, and recognized, in 1936, that the pattern of P (pressure) waves (the first to arrive after an earthquake) captured by seismological stations betrayed the presence of a solid body at the heart of the Earth's molten core. She described this discovery via what is probably the scientific paper with the shortest-ever title: *P′*. And, for such delicious brevity, one might forgive her any number of diplomatic lapses.

What is it like, this evolving solid core? Some idea of the rock is given by the nickel-iron meteorites that fall to Earth, though these were derived from the break-up of rather smaller bodies than the Earth. They are made of interlocking crystals of iron-nickel alloys, the two most common ones being the minerals kamacite and taenite. In structure and texture, the core may be anything but homogenous. Some scientists have theorized that this metal 'planet within a planet' might have its own surface tectonics, and may not have a smooth growing surface but rather a branching filigree of crystals growing out from the surface.

It would be a fine thing to observe such an inner planetary landscape, if our imagined probe could reach that far. However simple or complex its structure, though, it is geologically temporary. In something like a billion years from now, the core will have solidified completely, putting an end to our protective magnetosphere—and, in the process, to any surviving life on the planet. And, of course, it will put a stop to the rocks conjured up by those Earthly life-forms, which are of such dazzling variety and complexity that they deserve a chapter of their own.

Chapter 6
Living rocks, evolving rocks

Life makes up a very particular kind of planetary chemistry: complex, of a non-equilibrium state, and with a tendency, directly or indirectly, to affect the structure of the minerals that are around it and often formed within it. This gives very distinctive qualities to the rocks that are produced: qualities that we can now commonly recognize on Earth, and that we can hope to find—or at least to look for—as we explore other planets.

Beginnings

The oldest traces of life used to date back to 3.8 billion years ago, among the oldest rocks on Earth, themselves mangled scraps of strata deep within the cores of ancient continents. The rocks here include strongly metamorphosed mudstones that once formed on some ancient sea floor. Despite the heat and pressure they were subsequently subjected to, they are visibly dark, because they include traces of organic carbon. Now that in itself, of course, might have been formed without the helping hand of biology, simply through chemical reactions. A famous experiment was carried out in 1952, by the US scientists Stanley Miller and Howard Urey, where electric sparks were passed through a mixture of simple compounds such as water vapour, hydrogen, methane, and ammonia. The result was a variety of more complex organic substances, including amino acids: the compounds that are the building blocks of proteins.

The fossilized organic matter in these ancient rocks also shows a marked enrichment in the light carbon isotope, ^{12}C, and corresponding depletion in the heavy ^{13}C. This is a strong indication of life, as living organisms prefer to absorb ^{12}C. It is not a certain indicator of life at this time, though, because some chemical processes can be similarly selective of the carbon isotopes.

Nevertheless, based on such evidence, life, just possibly, might extend back yet farther, into the depths of the Hadean. A single zircon crystal 4.1 billion years old from Australia was recently found to contain tiny flecks of graphite. These flecks, analysed very carefully, also showed a marked preponderance of the light carbon isotope, hinting that life may have geologically arisen, or perhaps arrived, on Earth very shortly after its formation.

However ambiguous these early hints are, better preserved rocks from about 3.5 billion years ago provide more tangible evidence that, not only was life present then, but it directly controlled the formation of the rocks. These rocks are structures called *stromatolites*, which are limestones that show distinctively curved patterns of fine laminations (Figure 25). Stromatolites may be found today, albeit rarely. They are most common in marine conditions that are too extreme for metazoan life-forms such as worms and crustaceans, for example in Shark Bay in Australia, where the hot sun evaporates the shallow seawater sufficiently to make conditions too saline for most metazoans. The more hardy microbial communities, though, build sticky microbial mats, which trap layers of washed-in limy sediment. The microbes grow through this new layer to create another surface mat, to repeat the process anew, promoting the recrystallization of the calcium carbonate particles. Eventually, large mound-like structures grow, and breaking them across (the modern examples harden very

25. A stromatolite, from the Dalradian rocks of Scotland.

rapidly: it would be rather painful to try to kick one) will reveal an internally layered structure very similar to the ancient examples of Archean times—similar enough to leave little doubt that the ancient examples are microbially constructed rocks.

Today, stromatolites only thrive in places where creatures such as worms and snails are excluded—because those active and hungry creatures would soon make a meal of them. For the best part of three billion years, though, when life was essentially wholly microbial, they were common elements of the sea floor.

Iron rocks

Another form of rock formed in those ancient times—one that we now depend on to provide a key element of the structures that we build and the tools that we make—is iron. When the Earth's oceans first formed, they lay beneath an atmosphere that contained nitrogen, carbon dioxide, and methane, but no oxygen—hence the ocean waters were anoxic throughout. In such conditions, iron is in its reduced, ferrous (Fe^{2+}), state and can dissolve in water in large amounts. Those waters were soon teeming with life adapted to such conditions. As that life evolved, it chanced upon new chemical pathways to help supply it with energy and with the matter to build itself. One of these pathways—photosynthesis—uses the energy of the Sun to power chemical reactions, and using this the microbes split water and carbon dioxide to make the material of their bodies; one of the by-products of this process is iron oxide particles, which drift down to settle upon the sea floor.

Over more than two billion years or so, the ocean waters were purged of iron, which built up on the ocean floor, forming masses of iron ore far greater than any that have formed since. The resulting rocks, in which the bright red iron-rich laminae alternate with pale silica-rich layers, are termed *Banded Iron Formations* (*BIFs*). These BIFs are striking rocks—and utterly indispensable to modern human civilization.

One puzzling thing about them is that they started forming in the oceans about a billion years before there are clear signs that oxygen had yet entered the atmosphere and begun to oxidize and redden sediments on land. Perhaps the oceans allowed a little of the oxygen to leak out into the air; or, more probably, the microbes initially evolved *anoxygenic* photosynthesis, an alternative pathway that liberated not oxygen but hydrogen.

Microbial ghosts

Even where distinct stromatolites did not form, sticky microbial mats almost certainly spread far and wide over the Precambrian seafloors, playing a key role in determining their physical and chemical characteristics. After death and burial, the microbes would typically decay, leaving little trace of their former presence. However, tell-tale signs of their existence can be found where ancient sedimentary layers show signs of having behaved oddly.

Sands, for instance, differ from clays in being non-cohesive, and being transported, deposited, and re-eroded (say, by storm currents) as individual loose grains on a sea floor. When clay-free sandstones show signs of having acted cohesively, being ripped up and redeposited like sticky mud layers, or when they have developed wrinkly 'elephant-skin' textures from the passage of a water current overhead rather than being eroded, then it seems clear that something was holding those grains together. That 'something' is now widely interpreted as having been the microbial mats on those sea floors.

These microbial mats likely played a part in the preservation of the *Ediacara fauna*, those enigmatic, soft-bodied, metazoan organisms that appeared about 600 million years ago, in very latest Precambrian time, before disappearing when the mobile and aggressive new metazoans of the Cambrian took control of the sea floor. Some of these Ediacaran fossils show signs of preservation as a 'death mask', where their rotting remains were covered by rapidly growing microbial mats on which minerals precipitated to capture the impression of the organism even as the carcass decayed away.

These microbial 'slimeworlds' of the Precambrian may not have been as overtly complex as the metazoan organisms that came to share (and partly eat) their empire—but they possessed their own considerable sophistication, if modern microbial mats are anything to judge by. When enormously magnified by modern microscopes, they emerge as intricate as coral reefs, their complex structures comprising hundreds of co-operative microbe species that together exploit the nutrient gradients around the entire structure. They possess a kind of communication—termed 'quorum sensing'—to allow or deny entry to passing microbes, depending on whether they are identified as friend or foe. It's a phenomenon to consider next time a seemingly 'simple' fossilized microbial rock from the Precambrian comes into view.

The metazoan rock revolution

Something very remarkable started to happen about 550 million years ago. By the time it had finished some thirty million years later, the world had changed utterly. The empire of the microbes—and latterly that of their passive Ediacaran passengers—had given way to a world that we would find familiar, with worms wriggling, crustacean-like organisms scuttling along the sea bed and venturing up onto the beach, and corals, sea urchins, and sponges dotted about. The 'Cambrian Explosion' of metazoan organisms remains largely a mystery, but by the time it had finished, every major phylum that we see today had evolved and was living within the seas (though not yet on land). Even the vertebrates had arrived—although this slowest of evolutionary success stories was not to appear in force for another hundred million years. The consequences of this unique and unparalleled evolutionary radiation can be seen clearly in the rocks.

The most immediate effects had to do with the evolution of muscularity and active feeding, which included hunting. The new animals burrowed and chewed their way through the sea floor sediments, ripping apart the microbial mats and disturbing the delicate sedimentary layering that had been allowed to accumulate more or less undisturbed for over three billon years. The signs of this activity can be seen as bioturbation (Figure 26), which is biological disturbance of original sedimentary layering, which ranges from tube-like structures cutting across rock strata to a complete obliteration of primary sedimentary stratification. It is the occurrence of this character (in the form of the appearance of a particularly distinctive type of burrow) that was chosen to mark the beginning of the Cambrian Period—and indeed simultaneously the beginning of the Palaeozoic Era, which was to last for another three hundred million years, and of the Phanerozoic Eon, which persists today. Today, almost all of the strata on the sea floor are affected by burrowing and disturbance

26. Bioturbation: casts of burrows on the underside of a bed of sandstone (the influx of sand filled in the open burrow systems of a muddy sea floor). (Silurian of Aberystwyth, Wales.)

of the sediment, except for a few places, like the Black Sea, where the bottom waters are too oxygen-starved to allow animals to exist (and, as we have seen, the hostile hypersaline waters of Shark Bay in Australia).

There were other, more subtle, signs of the animal empire at work. One of the key factors allowing the spread of animals across the sea floor was the increased levels of oxygen in the atmosphere, and its spread through the ocean waters. However, in the Precambrian, these ocean waters would have been murky and turbid with the dead and decaying remains of planktonic algae that would drift in the water, using up oxygen as they decayed before they slowly sank to the sea floor. With the arrival of animals, this soup of decaying organic fragments began to be cleared away: partly by the action of planktonic organisms eating their way through this material and then packaging the digested remains into faecal pellets that rapidly fell through the water (to act in turn as food for bottom-living animals); and partly by the new filter-feeders on the sea floor—sponges and their ilk—that cleaned up the lower layers of ocean water. Thus, the new animal communities acted to modify the wider environment around them (mostly to their overall benefit)—and hence to alter the general characters of sedimentary rocks.

Many of the new animals had skeletons or armour of some sort—of calcium carbonate, phosphate, or silica—and so could be easily fossilized. The sudden appearance of fossils in strata at the beginning of the Cambrian—something remarked upon by Charles Darwin—and the successive evolution of new life-forms and new types of skeleton was, much later, exploited by humans who developed the science of *biostratigraphy*: the dating of rock layers based upon their fossil content. This science has now been refined to the point that time-slices often less than a million years in duration can be identified and traced around the world. This has allowed the history of the last half-billion years of our planet to be reconstructed in very great detail—far greater than is possible for the huge, but much less well-characterized, vistas of Precambrian time.

The reef-builders

The skeletons of marine animals can be used to serve a range of purposes, depending on the need of those making use of them: attack, defence, muscle attachments and levers, and platforms to help grow into the sunlight. In places, they pile up in such abundance that they can build entire rock layers—and those rock layers can in turn pile up in masses that can be several kilometres in thickness.

These layers can consist of masses of seashells washed together along some shoreline, which might eventually harden to form a layer of shelly limestone. However, some of the mightiest rock builders have combined to form constructions that are truly on a planetary scale, being up to several kilometres thick and hundreds—or even thousands—of kilometres in length. These are reefs. From the point of view of mariners, reefs are ship-threatening rock masses that are all too often in dangerously rough water. From the point of view of the reef-building organisms, though, they are a means to build a durable rocky platform, to stay within the warm, shallow, and sunlit waters in which they thrive.

Currently, the most visible rock-builders of this sort are modern corals (which are scleractinian corals, more commonly termed 'hexacorals'). These mostly colonial organisms secrete skeletons of calcium carbonate of a variety of different forms, from rounded boulder-like masses a few metres across, to delicate, fast-growing branching structures. The form partly depends on the species and partly on the local environment: the rougher and more wave-tossed the waters, the more rounded and robust the structures that are built by the coral animals. The outermost parts of the reefs, which take the brunt of the waves, are often built not of corals but of coralline algae, which produce even tougher masses of 'living rock'.

Such coral-algal growths today act as the complex framework and shelter for the extraordinary biological diversity of reefs, which rivals that of the tropical rainforests on land. A myriad species of mollusc, fish, worm, starfish, sea urchin, and other groups flourish and interact in the reef environment (though the 'reef' vistas gazed at by scuba divers are usually within the quiet lagoonal waters behind the reef itself). This biological richness is all the more extraordinary because it arises within nutrient-poor waters, being based on rapid biological recycling rather than massive nutrient input. Over-fertilize the waters of a reef, and fleshy algae (seaweeds) begin to grow profusely, to out-compete and eventually replace the corals, and this changes and impoverishes the whole reef ecosystem.

27. These jagged mountains built of massive, unstratified rock represent a ~200-million-year-old reef structure near to what is now the Gosausee lake, in Austria.

To a geologist, though, this marvellous reef biodiversity is simply a thin living skin upon rock masses (Figure 27) that may be tens, hundreds, or even thousands of metres thick. The thickness of these rock masses will depend on how old the reef is and on how much the sea floor beneath has subsided, with the corals always trying to grow upwards over the skeletons of their ancestors in an effort to stay in the sunlit shallows.

Reefs and reef-like limestone structures (often termed 'carbonate buildups') are common in the geological record of the past half billion years, and have been forming since complex calcium-carbonate-secreting organisms appeared. They have not all been based on corals, though. In the Cambrian, enigmatic tube-like organisms called archaeocyathids built reefs, as did, later in the Palaeozoic, another extinct group, the stromatoporoids, now thought to be forms of sponges. Other Palaeozoic reefs were built by extinct forms of coral (the rugose and tabulate corals), and late in the Palaeozoic a bizarre family of brachiopods (lamp-shells) called the richthofenids built tube-like shells that aggregated together to form reefs (in the Mesozoic Era a similar adaptation was made by molluscs called rudists).

There have also been 'reef gaps' in the geological record, usually following mass extinctions when the main reef-building organisms of the time were wiped out or greatly reduced in number; it usually took some several million years for newly evolved organisms to begin to reform the reef structures. Today, once more, corals are threatened by a combination of warming waters, ocean acidification, pollution by fertilizers washing off the land, and over-fishing (which removes the fish populations that, by grazing on seaweed, keep the reef structures healthy). Corals are particularly prone to acidification as their skeleton is made up of a crystalline form of calcium carbonate termed aragonite. This mineral is a little harder than the better known calcium carbonate mineral that is calcite, but it is more prone to dissolution as ocean pH falls. About a quarter of the world's reefs are now effectively dead; half are in a variable, often poor state of health; and only a quarter are healthy. The pressures on them show little sign of abating and hence humans may be producing yet another reef gap in Earth history, and stopping, for a few million years, the production of a beautiful and distinctive rock type.

Plankton rain

Reef limestones are essentially 'carbonate factories' that can only build limestone rock in shallow water. If the sea level rises too fast for the corals to catch up, then the reef structure may in effect be drowned—killed off as a rock-producer.

However, the tiny planktonic organisms that live in the surface oceans can form rock too, if their remains sink to the ocean floor in sufficient numbers to build rock strata in their own right. These kinds of rocks usually form far out in the ocean (where the rain of falling plankton skeletons is not too diluted by sand and mud washed in from the land).

In Palaeozoic times, and up to the late Jurassic, the main rock builders of this kind were the radiolaria, single-celled animals that secreted tiny, intricate skeletons of opaline silica. Their remains could build up on these distant ocean floors to form layers of radiolarite which, once buried and compacted, recrystallized to form hard, brittle layers of chert rock (in which tiny spherical 'ghosts' of the original radiolarians can sometimes be seen). Radiolarians, though, need a reasonable amount of nutrients to grow; in the nutrient-poor 'deserts' in the central ocean gyres they are scarce or absent, and all that accumulates on the sea floor is the finest and most far-travelled of wind-blown and volcanic dust, which accumulates very slowly as red and brown clays.

Ancient deep-ocean rocks of this kind are scarce because they, together with the ocean crust upon which they lay, have mostly been carried down subduction zones at ocean trenches, eventually being absorbed into the Earth's mantle. It is only thin slivers that have been scraped off in the subduction process and plastered onto the facing continental margin as ophiolites, which now remain for us to study.

A revolution in the plankton, though, took place in late Jurassic times. Calcareous, planktonic organisms evolved and radiated out into the oceans. They were dominated by the coccolithophores—single-celled algae that secrete a skeleton made of complex microscopic shield-like structures (coccoliths) of calcium carbonate—together with planktonic foraminifera, amoeba-like organisms with tiny, chambered, calcium carbonate shells. These could live in more nutrient-poor waters than the radiolarians and so, from late Jurassic times until today, they have rained down as oozes from the central expanses of the oceans.

Not all of the skeletons arrive on the ocean floor, though. Where the very deep waters are too acidic, being charged with too much carbon dioxide from respiration and decay, these skeletons dissolve before they reach the ocean floor. Hence, there is a kind of 'snowline' known as the *carbonate compensation depth* (*CCD*) that today is typically at depths of between 3 and 4 km, above which they can carpet the ocean floor (on the tops and flanks of sea mounts, say) and below which it is only the less easily soluble, silica-based plankton skeletons or deep ocean clays that accumulate.

The oozes on the very tops of these undersea mountains may be thickly sprinkled with the skeletons of pteropods (pelagic molluscs often called 'sea butterflies'). These skeletons, like those of modern corals (discussed earlier), are made of the relatively soluble form of calcium carbonate, aragonite, and hence the sinking skeletons, on their journey downward to the ocean depths, dissolve relatively quickly, while the sinking skeletons of calcitic organisms such as the foraminifers and the coccolithophores reach greater depths before they, too, are dissolved. Pteropods, therefore, are in the front line of organisms threatened by contemporary human-driven acidification of the oceans. Their skeletons are already perceptibly thinner than in pre-industrial times, and the near-future ending of pteropod limestone production now seems more likely than not.

This kind of deep ocean limestone has its own ancient history: the iconic chalk deposits of the Cretaceous Period (Figure 28) are rocks of this kind that formed globally in the greenhouse world of this time, when virtually all the icecaps had melted, in effect spreading deep-water 'oceanic' conditions over much of the continents. A little later, at the end of the Paleocene Epoch, fifty-five million years ago, the CCD, the oceanic 'snowline' of calcium carbonate oozes, rose worldwide by a kilometre or so, as the natural release of a few trillion tons of carbon (in the form of carbon dioxide and/or methane) into the ocean/atmosphere

28. Chalk of Cretaceous age in Italy. The unit of darker strata represents an oceanic anoxic event, when the sea floor became starved of oxygen, and therefore enriched in organic matter.

system acidified the oceans. This was a kind of natural precursor of humanity's own ongoing experiment in changing the pattern of deep-water limestone production.

The plankton that form these oozes and, ultimately, chalk rocks, are sensitive to a wide range of conditions around them. They form in greater or lesser numbers or types as the climate changes and the waters that bathe them become warmer or cooler, or a little less or more nutrient-rich. They also register subtle features of ocean chemistry, such the relative proportions of oxygen isotopes in the water molecules (that change as polar ice grows or melts, drawing water out of the oceans or flooding them with meltwater). Hence, analysis of these deep sea deposits has been crucial in deciphering the twists and turns of the Earth's climate over the past hundred million years and more: it has been one of the main achievements of the deep sea drilling programme—a true, yet largely unsung, revolution in the Earth sciences.

Rock-forming life on land

Life has been essentially a marine phenomenon since it first originated, throughout the almost eternal duration of the Precambrian slimeworld and for the first part of the Palaeozoic Era, when trilobites, corals, molluscs, and other metazoans flourished in the sea. Throughout this long span of time, the land remained a lifeless desert: or, perhaps more truthfully, almost a lifeless desert. There are hints that some forms of algae, and maybe even a few simple plants resembling modern liverworts, might have clung to some low-lying damp patches of the landscape.

It is only by midway in the Palaeozoic Era, in the Silurian Period, that simple fossil plants—mostly simple branching green stems a few centimetres high—began to root themselves into the land, and also to be preserved into the fossil record. And it was the succeeding Devonian Period, which began about 400 million years ago, that saw these pioneers evolve to take over the landscape to the extent that they became rock-builders in their own right. From then on, they became powerful controlling and modifying agents of most rock strata that formed on land.

During the Devonian Period plants reached sufficient abundance to form the first peat bogs. One of these was preserved in astonishing detail by the silica-rich, mineralizing waters emanating from a volcano near what is now Rhynie, in Scotland. The plant stems are preserved in cellular detail—together with an ecosystem of arthropods that followed the plants up onto the land. These include early mites, springtails, and harvestmen (puncture marks from herbivorous arthropods are preserved, too, on some stems).

Later in the Devonian the first forests appeared, the oldest known fossil example being found in the rocks of what is now Gilboa, New York, where trees at least 8 metres high grew on an ancient coastal plain. It was in the succeeding Carboniferous Period that the forests grew to a scale comparable to the forests of today. However, these forests were populated not by pine and oak but by such plants as giant horsetails and ferns, between which flew dragonflies with wingspans reaching 30 cm, while early amphibians crawled and swam through the swampy morass below. By a geological coincidence, these Carboniferous forests coincided with a time when such swamp and delta systems extended widely around the world in equatorial regions, and so they were buried to form gigantic coal deposits, which we mine and burn in equally giant amounts today.

The plants not only provided the coal seams and the beautiful fossilized impressions of ferns that can be found in the intervening mudstones, their roots anchored the loose sediment on the surface, turning it into proper soils, which were then enriched with their decaying remains. The fossilized remains of such soils (*palaeosols*) date from this time. Where rivers flowed, the plants and their extensive root systems anchored the banks. This meant that the river channels could no longer split and change

29. A piece of the Ludlow Bone Bed—about 400 million years old. The large fragments are fin-support spines of early fish, while the particles that look like sand grains are tiny fish scales. (About 2 cm across.)

direction so easily. River-laid strata from before the time of land plants typically show the sand- and gravel-dominated pattern of braided river systems, while those afterwards are commonly of a meandering pattern, in which the sandstones originated as sandbars on the meander bends while thick mudstone layers represent the vegetated floodplains.

Bone beds

The newly evolved animals of the vegetated landscapes of the past 400 million years are usually rarities as fossils, occasionally chanced upon by the lucky geologist—often with great celebration afterwards. However, now and again they accumulate in larger concentrations, to form deposits that are of great interest to the palaeontologist—but that can also be of great use to hungry humans.

These are bone beds, which typically form where disarticulated animal bones, teeth, and scales are washed by river currents (or coastal currents, if washed into the sea) and act as pebbles and cobbles, separated from the surrounding sediment by the winnowing action of the water. Thus concentrated, they can form distinctive, if usually thin, layers that can act as valuable marker beds where geological strata are mapped.

A remarkable early example is found around the market town of Ludlow, in England. It is of late Silurian age, and so dates from the very beginnings of life's colonization of the land. It is usually only an inch or so thick, and looks a little like coarse gingerbread, but it is made up of the scales and spines of early armoured fish and crustaceans, winnowed by coastal currents along a coastline (Figure 29). Even more remarkably, the washed-in fragments include—*exceedingly* rarely—the remains of some of the first arthropods—arachnids and myriapods—that scuttled across the landscape.

Bone beds turn up sporadically elsewhere, with spectacular examples in the Dinosaur National Monument in the USA and in Mongolia's Gobi desert. In eastern England there are several within the early Cretaceous strata, which include, as well as bones, structures termed *coprolites*, some of which represent the petrified faeces of dinosaurs or marine reptiles. In the middle of the 19th century, when England's population was booming and the farmers were struggling to feed everybody, it was discovered that these fragments (which, being bone, are phosphate-rich) made a superb fertilizer when crushed and acid-treated. A thriving and highly profitable industry formed to quarry away these 'coprolite beds'.

Some considerable figures were involved in this industry. John Henslow, Charles Darwin's beloved mentor of his time at Cambridge, seems to have first encouraged the farmers of eastern England to use such fossil manure. William Buckland also became involved. An extraordinary combination of early savant of geology at Oxford and Dean of Westminster, he was the first to scientifically describe a dinosaur (*Megalosaurus*); carried out his fieldwork in academic gown; reputedly ate his way through the entire animal kingdom; and coined the term 'coprolite', using these petrified droppings to help reconstruct the ecology of ancient animals. Later, he energetically collaborated with the celebrated German chemist Justus Liebig (who had worked out how to chemically treat these fossil phosphates to make fertilizer) to show how they could be used by agriculturalists, once demonstrating their efficacy by exhibiting, in Oxford's Ashmolean Museum, a turnip, a yard in circumference, that he had grown with such prehistoric assistance.

It is related strata (geologically rare phosphate-rich deposits, usually biologically formed) that are still a mainstay—if a rapidly depleting one—of modern agriculture. In a very real sense, these particular rocks are keeping us all alive.

Chapter 7
Rocks on other planets

Rocks are not just an Earthly phenomenon. They make up the surfaces of some of the planets of the solar system, and—in one form or another—of all of the many moons and innumerable comets and other small objects that orbit out to great distances from the Sun. They likely also lie deep beneath the immensely thick fluid envelopes of the gas giants. Scientists have now imaged, and physically and chemically sensed, many of these rock surfaces—they have even taken (but rarely retrieved) samples from a few of them. These extraterrestrial rocks show an extraordinary variety that continue to surprise us with each now discovery. The beginnings of exploration of exoplanets suggests, too, that an even wider variety of rock formations likely exists on star systems other than ours.

Mercury rocks

The planet Mercury is so close to the Sun that direct observation of its surface has been difficult to achieve against the glare of intense radiation streaming past it. Mercury is very dense, mostly comprising an iron-nickel core. The rocky mantle above is comparatively thin—perhaps because most of it was stripped away in an early planetary collision—and so the form of the crust, only dimly perceived through telescopes on Earth, was a major question in the minds of planetary scientists as the MESSENGER spacecraft approached in 2011 to take a close look.

The heavily cratered surface superficially resembles that of the Moon, and spectral analysis shows that it is largely made up of basalt which is compositionally broadly like that of basalts on Earth, underlining the pervasiveness of the partial melting process which produces this rock type on planetary surfaces, even on planets of very different internal anatomy. The heaviest bombardment took place some four billion years ago, and perhaps helped triggered further lava flows which infilled much of the rugged impact-generated early topography. Surprisingly, despite the proximity of the huge iron core (which seems to have a kind of sandwich structure, with solid inner and outer parts separating a melt layer) these basalts were seen to be rather iron-poor and magnesium-rich. They flooded out as hot, runny lavas to make up a smooth surface that was then bombarded by subsequent meteorites, and that also contracted overall to produce wrinkles and fault scarps that are still detectable today. Yet more surprisingly, even so close to the Sun, solid deposits of a mineral thought to be water ice have been detected in some sheltered areas within the deepest craters near the poles.

Venusian rocks

The rocky surface of Venus was for a long time a mystery, completely shrouded by its permanent cover of bright opaque clouds. The landscape beneath was usually inferred to be draped in tropical-style jungle, which could (in a thousand science-fiction novellas) be inhabited by all manner of Venusian beings, to be fought—and occasionally wooed—by intrepid Earthling spacemen. That exotic vision was dashed and replaced by a far more hellish one, as the Russian *Venera* spacecraft became the first to penetrate the cloud cover. They revealed a surface temperature of over 400°C and a near-complete absence of water—boiled away in a runaway greenhouse effect caused by the thick carbon dioxide atmosphere, then stripped off into space by the solar wind—while the clouds themselves are of concentrated sulphuric acid droplets.

What lies beneath is a volcanic landscape, which has now been mapped by cloud-penetrating radar from orbiting satellites (Figure 30). There are lava channels that can be many hundreds of kilometres long; lava tubes ten times the size of those on Earth; volcanoes that include swarms of 'pancake' forms, like small very flat shield volcanoes; and 'coronae', enormous crown-shaped forms inferred to form as plumes of Venusian mantle produce bulges on the crust, generating lavas which spill out along the bulge margins, before the central parts sag down as the magma chambers are emptied. Spectral analysis of surface rock is difficult through the thick atmosphere, but the overall form of the volcanic features and the few analyses (briefly) made by landing craft suggest a surface largely formed by the eruption of hot, highly mobile basalts.

The volcanic surface is here and there cratered by incoming meteorites, but these are curiously evenly distributed and suggest that the rocky surface of the planet is all more or less the same age: about half a billion years old or less. It may be that the volcanism is very evenly distributed around the planet to achieve this effect, but a more dramatic hypothesis to explain this pattern is that the bulk of the volcanism is concentrated in brief planet-wide bursts of 'resurfacing', as magma pours out across the planet. Whichever explanation is true, what Venus clearly lacks is a plate tectonic mechanism like the Earth's, even though it is much the same size and has a similar overall composition. The difference is probably due to the lack of water—and oceans—on Venus, which on Earth function as a lubricating fluid to ease the friction-ridden passage of the downgoing plates along subduction zones.

While about three-quarters of Venus's surface is of solid rock, there are some sedimentary deposits derived from the fragments

30. A three-dimensional perspective image of Maat Mons, part of the Venusian volcanic landscape.

generated by meteorite impacts and by any explosive volcanism, and shaped by winds in the dense atmosphere into dune-like structures. There is probably some surface alteration of the rock material too, although this will be very different to the water-mediated chemical alteration of rock on Earth. On Venus the combination of carbon dioxide, sulphur dioxide, and concentrated sulphuric acid should produce carbonates, quartz, and sulphates as secondary minerals—which perhaps might cement the grains, here and there, into a kind of Venusian sandstone.

One wonders whether, buried beneath those abundant volcanic rocks, something might remain of an early Venusian crust, which might include traces of former oceans—and possibly even of the beginnings of some kind of early life—on this now emphatically dead planet. Locating such a geological relic one day—when human space exploration efforts are *much* more advanced—would be a major, if perhaps sobering, step in understanding the history of our own star system.

The Moon

Our own Moon is another dry, rocky body that is close enough to have been visited and sampled—albeit that sampling was carried out almost half a century ago, and those few extra-terrestrial journeys have not been repeated since. Here the surface shows contrasting terrains, with ancient, meteorite-pounded highlands and smoother plains (the 'seas') of dark basaltic lava that later (mostly between three and 3.5 billion years ago) oozed out to fill the low ground. The highlands are impact breccias of a rock that was originally a light-coloured, low-density anorthosite rock made up of the calcium-rich feldspar anorthite. This ancient rock is probably 'primordial', representing floating masses of the first crystals to form on the original magma ocean.

Detailed chemical analysis of all of the Moon rocks suggests that they are very closely related to Earth rocks: the dynamics of the Earth–Moon system suggests that the Moon was formed when the proto-Earth, soon after its formation, was struck a glancing blow by a Mars-sized planet (Theia) which was completely obliterated (see Chapter 2). Part of Theia merged with the proto-Earth (itself completely melted by the impact) to form our present planet, while the debris flung out coalesced to form the Moon. The very close similarity of Earth and Moon rock chemistry has been a problem with the impact theory (Mars, for instance, has very different isotopic chemistry). Various explanations have been put forward to square this particular circle. A recent one suggests that proto-Earth and Theia formed in very similar orbits, to give them similar chemistry—and that would of course mean that they were fated to collide sooner or later.

Moon rocks on Earth include 381 kg collected during six Apollo missions between 1969 and 1972, and a third of a kilogram of lunar soil collected and brought to Earth by three Soviet-era Luna landers between 1970 and 1976. Some the Apollo material was given as goodwill gifts to different countries and US states by President Richard Nixon. Many of the gifts subsequently had chequered histories, include loss, theft, and deception. In one case, the recovery of one such Moon rock gift was via an elaborate 'sting' operation co-ordinated by NASA and the US customs and postal services, and involving bait of $5 million (put up by the then presidential candidate H. Ross Perot).

Such skulduggery is not absolutely necessary in order to lay one's hands on some Moon rock. Over a hundred meteorites have been identified as coming from the Moon, having similar chemistries to the Apollo- and Luna-derived samples. Some date back 4.5 million years and so are older than any surviving Earth rock. These lunar meteorites are highly sought after by collectors, so if you do wish to possess one it does help to be rich.

Rocks of the red planet

Mars is not quite so hostile a place (to humans or human-made machines) as Venus, and so it has been explored much more closely. It is not exactly habitable—it has a very tenuous carbon dioxide atmosphere and is bitterly cold by Earth standards, ranging from a few degrees below zero at its Equator to about –180°C at the poles. And it may be—may always have been—completely dead (although there is avid interest, and much research, into this question).

Mars has both igneous and sedimentary rocks, though, as far as we know, not much in the way of metamorphic rocks (except for those which have been shock-altered by incoming meteorites). Like our Moon, it is divided into ancient meteorite-battered highlands and more recent lowlands, although the lowlands here seem to be mainly sedimentary, rather than made of volcanic lavas. And the word 'recent', for the most part, is applied here in relative terms. The large features of Mars's rock structure, both volcanic and sedimentary, were established in its first one to two billion years—during Earth's Hadean and early Archean eons—and most subsequent activity took the form of surface modification by ice (largely as permafrost) and the winds, which still blow powerfully even in such a thin atmosphere.

The largest feature of all, the hemispheric separation of the planet into the southern highlands and the lowlands to the north, might—it is controversial—reflect a massive early impact that almost shattered the young planet. Subsequent history saw the continued battering of the higher ground and filling in of the enormous basin to the north, probably by a combination of lava flows and sediments. The volcanic foundations of the ancient battered crust of the highlands seem to be mainly basaltic, to maintain this most typical planetary pattern of rock production. However, spectral analysis from satellites has also recognized patches of more silica-rich rock, including those broadly thought to be of granitic composition. This early volcanism may have included explosive volcanism, spreading volcanic ash far and wide, as well as the quiet effusion of lavas. Some of the ancient craters have been interpreted as volcanic rather than impact-related, and compared with the kind of large calderas on Earth that have generated powerful and destructive eruptions. The release of gas from magma—a key driver of such eruptions—may have been aided by the low pressures on this small planet.

There are tiny rock samples of these Mars rocks on Earth—a little over a dozen meteorites are known to have come from this planet because features of their chemistry, including the composition of trapped gases, match nothing on Earth but closely resemble those of rocks and atmosphere analysed on Mars. In composition they are igneous rocks broadly of basaltic type, but with varying proportions of component minerals such as olivine and pyroxene.

They range in original crystallization age from one to four billion years old, though some also include evidence of other events in their history, such as contact they had with water, or when they were shock-blasted off the Martian surface by meteorite impact.

One of these Martian rocks, the 'Allan Hills 84001' meteorite, discovered in Antarctica in 1984, is famous (or perhaps notorious) for bearing possible evidence of life, including exceedingly tiny microstructures and mineral particles that were originally interpreted as possible remains of Martian microbes (see Chapter 3). Subsequent research cast doubt on this interpretation, suggesting that these 'microbes' could as easily have been formed by purely chemical means, which currently seems the most likely interpretation. It is still an open question whether any life exists, or ever existed, on Mars.

A late stage in this volcanism is represented by the enormous spectacular volcanoes of the 'Tharsis Bulge'. This is a raised area of crust some 5,000 km across on which are dotted some of the largest volcanoes in the solar system, including the mighty Olympus Mons; at nearly 25 km high it is almost three times the height of Earth's biggest volcano, Mauna Loa on Hawai'i (measuring the latter from the bottom of the ocean to the tip of the crater). Like Mauna Loa, it is essentially a shield volcano, made up of innumerable flows of basalt. It may not be entirely extinct. The scarcity of craters on its flanks suggest that part, at least, of the edifice was constructed in the past hundred million years, with some lavas being perhaps only a couple of million years old.

The primary volcanic rocks of Mars were, in that tumultuous first billion years, eroded by winds and rain, as well as by the bludgeoning meteorites. The action of wind has long been clear—indeed, ever since the first Mariner spacecraft went into orbit in 1971 only for its view to be blocked by the opaque clouds of a planet-wide dust storm. Only when those clouds settled, several weeks later, could the first pictures of the Mars surface be taken. Successively more detailed pictures have showed mobile sand dunes that look surprisingly like those of Earth—surprisingly, because on Mars gravity is only about one-third, and the atmosphere one-hundredth the density, of our planet. This aridity will have been a constant of Mars for billions of years, and so the strata of Mars will include a good proportion of sandstones made of eroded grains of basalt, sculpted into ancient wind-dunes.

Many of the surface rocks and boulders are clearly wind-sculpted and smoothed. There are rock structures like the yardangs of Earth: parallel, metre-scale, ridge and furrow systems, sand-blasted into solid rock. There are others that are today vanishingly rare on Earth (though they may have been common in Hadean and Archean times): pedastel craters, where the wind has picked away at the softer sediment at the edges of the outflung blankets of ejected rubble to create cliff-like margins.

The role of water in shaping the rock strata of Mars has been much harder to work out. The science here has had a notoriously chequered history, since Giovanni Schiaparelli published a detailed map showing a network of canals he claimed to have seen by telescope in 1877, with Percival Lowell, a wealthy US industrialist turned astronomer, elaborating this vision. In Lowell's 1895 book, *Mars*, he listed 183 named canals, from *Acalandrus* to *Xanthus*, complete with the number of observations made of each (totalling 3,240) and persuasive discussion (the water was drawn from the icecaps of that planet, he said, and hence fresh). Alas, none of this existed: it was all optical illusion, from beginning to end.

The modern satellite images clearly show canyons and channels, in places arranged in what look like distributary systems, but these are quite unlike Schiaparelli's and Lowell's imagined regular network. Today, little or no water can flow along them, given the frigid temperatures; and, in the distant past, the Sun was likely almost a fifth fainter than it is today. So, to have temperatures warm enough for sustained flowing water would have needed a much thicker, heat-trapping atmosphere. Hence, it was suggested that canyons may have been cut by dry suspensions of sediment in air, akin to dense volcanic currents. Alternatively, if there were floods of water, these may have been produced as giant meteorite impacts melted permafrost to generate catastrophic floods of scalding meltwater. These floods would have rapidly carved landscapes and left thick beds of sediment before the planetary surface froze again.

However, the detailed observations of the past few years—particularly from the NASA rovers trundling across the Martian landscapes—have slowly, painstakingly accumulated evidence that the Martian surface did once enjoy more or less sustained climate episodes mild enough to generate rivers and lakes in which strata (Figure 31) could be slowly deposited, and in which rock weathering processes could make significant amounts of clay deposits.

One rock type, though, is notable for its absence: limestone. In this, Mars is out of kilter when compared with other rocky planets.

31. Mars strata at the base of Mount Sharp within Gale crater.

Earth has a lot of limestone and strata-bound hydrocarbon deposits, and a little carbon dioxide in its atmosphere. Venus has no limestone to speak of, but it has an equivalent mass to Earth's carbon stores held as carbon dioxide in its crushingly thick atmosphere. Mars has a carbon-dioxide-dominated atmosphere, true—but a very tenuous one, and, except for thin layers of solid carbon dioxide in its icecaps, nothing significant in the way of limestone or hydrocarbon deposits that have yet been detected. So where has this rock gone?

Most likely its components have streamed away into outer space. The growing evidence for a warmer, wetter early Mars at a time of a weaker Sun is a strong indication of a thicker atmosphere, with carbon dioxide likely being a key component. And measurements of the Martian atmosphere today show that small amounts (about a hundred grams a second) are being stripped away by the solar wind. A hundred grams does not sound much, but maintain that rate over billions of years—with no protection for most of that time from any sort of magnetic field—and it can build up to carbon loss, and the absence of limestone rock—on a planetary scale.

And finally, before we leave Mars—why are the rocks of the red planet red? The redness, for a start, is only skin-deep—it is strictly a surface phenomenon. It does seem to be due to an oxide (finely divided haematite (Fe_2O_3) is a strong red colour). Any oxygen present in the atmosphere has only ever been present in exceedingly small amounts—but over geological ages that could have been enough to give the planet's surface its distinctive hue. It might have come from photodissociation of tiny amounts of water vapour; or perhaps the oxidant was hydrogen peroxide; or perhaps water simply reacted with reduced iron in the volcanic rocks to give iron hydroxide, which subsequently decomposed to an oxide. It is a small puzzle, among the many mysteries of this fourth planet out from the Sun.

Rocks of the gas giants

It would be hard for a geologist, armed with hammer and even the best of space-suits, to obtain anything in the way of a goodly lump of rock from Jupiter or Saturn. These are the gas giants, and echoes of the ancient weather systems that swirl across their surfaces likely penetrate thousands of kilometres into their fluid interiors, which are made of increasingly tightly compressed hydrogen and helium, with some methane, water vapour, and ammonia.

Nevertheless, deep in their interiors there might be solid cores of rock and metal—or at least they were likely originally, when these planets accreted. Models of Jupiter, for instance, suggest a rocky core of some sort surrounded by hot metallic hydrogen. However, it cannot be excluded that this core has been dispersed and redistributed to higher levels in the planet by currents in the metallic hydrogen. Jupiter might therefore be a rock-free planet, even at depth.

Uranus and Neptune are a little different: smaller than Jupiter and Saturn, they are ice giants rather than gas giants, being mostly made of ammonia- and methane-laced ice beneath their dense atmospheres, with rocky cores at the centre. The nature of the rocks at these depths may be truly fantastical. High-pressure experiments suggest that the methane in the 'ice' mantle of both planets (at this depth it is in reality a hot dense fluid) may break up to form a rain of crystalline diamond, which would fall into an ocean of liquid diamond around the core, on which might float solid 'diamond-bergs'.

The distant moons

While the distant planets of the solar system only provide tantalizing hints of the rock that lies deep inside them, there is much more tangible evidence to be found on the moons that circle them. Once thought to be rather dull and uniform objects, the spacecraft launched to study them found an astonishing diversity of form—and of geology, too.

Jupiter has a dozen or so moons, and the nearest to the planet, Io, originally discovered by Galileo in 1610 CE, possesses dramatic geology, the discovery of which represents a lovely example of scientific prediction and testing. A paper published a little before the Voyager 1 spacecraft flyby predicted that the moon's interior should be sufficiently stirred by the tidal forces generated by its interaction with Jupiter to produce volcanism. The early pictures revealed sulphur-rich plumes reaching up to 500 km into its thin atmosphere. The scientist Linda Morabito, who had survived a traumatic and abusive childhood to become a member of the famed Jet Propulsion Laboratory in California, showed that these were volcanic, and that Io is the most volcanically active body in the solar system. Io's constant eruptions are driven by tidal forces which are strong enough to produce a substantial magma ocean below Io's crust. Initially, the eruptions were thought to be largely of sulphur-rich material. However, it turned out, from observations made by the Galileo space mission, that the fountains of ash and lava were derived from rock magma of generally basaltic composition; sulphur is also present, and is largely responsible for the varieties of colour seen on this moon's surface.

Farther out from Jupiter, Europa is surfaced by the rock type that is to be expected in these distant regions—water ice. The ice overlies a salty water ocean perhaps 100 km deep. Its surface is overall smooth and there are few craters, suggesting that it is geologically young. The surface ice is criss-crossed by complex networks of fractures, some in more or less organized patterns that have even been compared to those of the plate tectonics of Earth. The brittle surface ice might be brought to the surface or carried into the interior by warmer convecting ice below, which is set into motion by the same tidal forces that drive Io's volcanoes. Other regions suggest a jumbled mass of blocks frozen together, and these might lie atop subcrustal water bodies. The ocean that lies below seems to be salty, as salts have been detected on this moon's surface. These salts may have been derived from exchange with the silicate rock and iron core that forms the ocean floor. Europa is a complex and interesting body—and a prime scientific target in the search for extraterrestrial life.

The moons of Saturn are dominated by one large body—Titan, which also has a thick water ice shell serving as crust. Titan has a dense nitrogen atmosphere with weather dominated by clouds and rain of liquid hydrocarbons, notably methane and ethane. These feed rivers that flow into hydrocarbon seas that in places—like the Kraken Mare—may be over 1,000 km across. The rivers and weather systems erode the ice into grains, pebbles, and boulders, some of which were glimpsed by the Huygens landing craft that settled on the surface of Titan in 2005. There are also surface dune systems (Figure 32) shaped by the local winds, where the 'sand' particles are made of ice or organic particles (the latter polymerized from hydrocarbon aerosols in

32. Wind-blown dunes on Titan, viewed from orbit by the Cassini satellite.

the nitrogen-based atmosphere) or both. One might expect, therefore, that there will be strata of ice sandstones and ice conglomerates, subsurface examples of which may serve as extensive hydrocarbon reservoirs. From its surface to its subterranean regions, then, Titan would be an oil geologist's playground. It may not be the easiest place to work, however, given the temperatures of –180°C; the intermittent hydrocarbon downpours that cause flash floods in river valleys; and the sticky, soot-like particles that may be the cause of the scattered dark patches on the moon's surface. When astronauts finally arrive, they are likely to find the terrain to be a messy and uncomfortable place to travel across and to work on.

Beneath Titan's ice shell there are more oceans, this time of water mixed with ammonia, perhaps a couple of hundred kilometres below the icy crust, lying above the silicate bedrock of the planet. It is yet another of the habitats that, scientists think, might possibly harbour life in the solar system.

Ice—often with subcrustal water oceans—dominates other moons. Callisto, another satellite of Jupiter, is another body with an ice shell that locally fractures, releasing geysers of salty water. Ice, too, makes up most of the material of the distant comets.

As more distant and even colder regions are approached, the nature of the rock changes. That celebrated dwarf planet, Pluto, was approached in 2015 by the New Horizons satellite. There was time to take only hurried pictures as the spacecraft sped by (and then it took many months for that data to be transmitted, exceedingly slowly, back to Earth). Nevertheless, these pictures showed an extraordinary, and quite unexpected, rocky landscape (Figure 33). The most rugged topography includes mountains of hard, brittle water ice, though quite how that topography was made remains largely a mystery. Some might have been made by 'cryovolcanism', as warmer, more mobile ice from the depths rose to the surface. It is so cold that much of the solid surface

33. Rock structures on Pluto: the mysterious 'dragon scale' terrain.

comprises smooth plains of nitrogen ice, with enigmatic, regular fracture patterns, and there are also mysterious 'dragon skin' landscapes of ridges and hollows. In these cold and distant regions, we are in entirely different realms of rock, the dynamics and processes of which will puzzle planetary scientists for a long time to come.

The rock of distant planets

Farther out even than Pluto, there are the distant, ultracold, thinly populated regions of the Kuiper Belt and then the Oort Cloud, where blocks of rock and ice—comets—orbit around the Sun, some occasionally approaching close enough to the Sun for volatiles to stream from them in classic coma form. Some of the bodies of the Kuiper Belt that are big enough to be observed are behaving strangely, as if their orbits were being perturbed by a much larger, though more distant body. Hence, there are suspicions of a 'Planet Nine', estimated to be several times the size of Earth, orbiting somewhere at the far fringes of our solar system. If it is ever observed in detail, its own frigid rocky surface will likely provide as many surprises as did that of Pluto.

Planets of other star systems

The mysterious, thinly populated Oort Cloud may extend for as much as two light-years from the Sun, thus reaching to half the distance to the Sun's nearest neighbour, the star Proxima Centauri.

Beyond that, there are other star systems and other planets. Humans have long known, and charted, the luminous stars, but the faint planets that orbit them were first detected in 1992, though now a few thousand of these extrasolar planets, or *exoplanets*, are known—or suspected. They are mainly recognized by the effect of their gravitational pull on their parent star, or by the way that they make the light from that star dim a fraction as they pass in front of it. Few have been directly imaged, so strong is the stellar glare surrounding them.

Our sampling of exoplanets is heavily biased towards those that are large and close to their star; it is much harder to detect the small, distant ones—and the moons that likely rotate around them are still quite out of view. Nevertheless, it is clear that our type of well-ordered solar system, of well-spaced planets with neatly near-circular orbits (and the rock structures that reflect that) may be a cosmic rarity. The haul of exoplanets is currently dominated by 'super-Earths' several times the size of our planet and 'hot Jupiters', gas giants that pass close to their star and so have roastingly hot planetary surfaces and often freakishly eccentric orbits. Relatively few have been found with, say, the conditions at their surface that may allow liquid water—and so the range of sedimentary rocks that we see on Earth. Phenomena such as volcanism can only be speculated upon (or perhaps modelled) on these very alien planets—though most of them will need some kind of heat release mechanism, and some may well have evolved mechanisms even more bizarre than Venus's 'resurfacing' or Earth's plate tectonics. The next generation of telescopes might allow us to glimpse some of these things. Planets, we now know, are very common in our galaxy (and hence likely equally so in other galaxies). Their rock formations will certainly include things to marvel at.

Chapter 8
Human-made rocks

Rocks are made out of minerals, and those minerals are not a constant of the universe. A little like biological organisms, they have evolved and diversified through time. As the minerals have evolved, so have the rocks that they make up. We have seen in the preceding chapters some of the ways in which rock units have changed through geological time. However, as regards both rocks and minerals—and the structures made of them—we seem to be entering a remarkable new phase of history. We do not know where this phase is leading to, or when or how it will end. But it is already clear that it is of planetary—and perhaps wider—importance, which we shall explore in this final chapter.

The evolution of minerals

The pattern of evolution of minerals was vividly outlined by Robert Hazen and his colleagues in what is now a classic paper published in 2008. They noted that in the depths of outer space, interstellar dust, as analysed by the astronomers' spectroscopes, seems to be built of only about a dozen minerals (see Chapter 1): various nitrides, carbides and oxides, diamond—and, of course, ice. Their component elements were forged in supernova explosions, and these minerals condensed among the matter and radiation that streamed out from these stellar outbursts.

As clouds of dust happened to swirl more densely, collapsing under their own weight, new star systems were born. The gas and dust whirling around a new star—our own infant Sun, for example—coalesced to form asteroid- and comet-sized lumps, which further aggregated into planetesimals. New minerals formed, which one can find in meteorites, and of which about 250 have been recognized so far.

Planetesimals began to aggregate into planets, with core, mantle, and crust, and the number of minerals on the new Earth rose to about 500 (while the smaller, largely dry Moon has about 350). Plate tectonics began, with its attendant processes of subduction, mountain building, and metamorphism. The number of minerals rose to about 1,500 on a planet that may still have been biologically dead.

The origin and spread of life at first did little to increase the number of mineral species, but once oxygen-producing photosynthesis started, then there was a great leap in mineral diversity as, for each mineral, various forms of oxide and hydroxide could crystallize. After this step, about two and a half billion years ago, there were over 4,000 minerals, most of them vanishingly rare. Since then, there may have been a slight increase in their numbers, associated with such events as the appearance and radiation of metazoan animals and plants (though these largely made elaborate arrangements of existing minerals within their skeletons, rather than creating new ones).

Then came modern *Homo sapiens*, and the nature of Earth's minerals changed (Figure 34).

Human-made minerals

Humans have begun to modify the chemistry and mineralogy of the Earth's surface, and this has included the manufacture of many new types of mineral. These, though, are minerals in

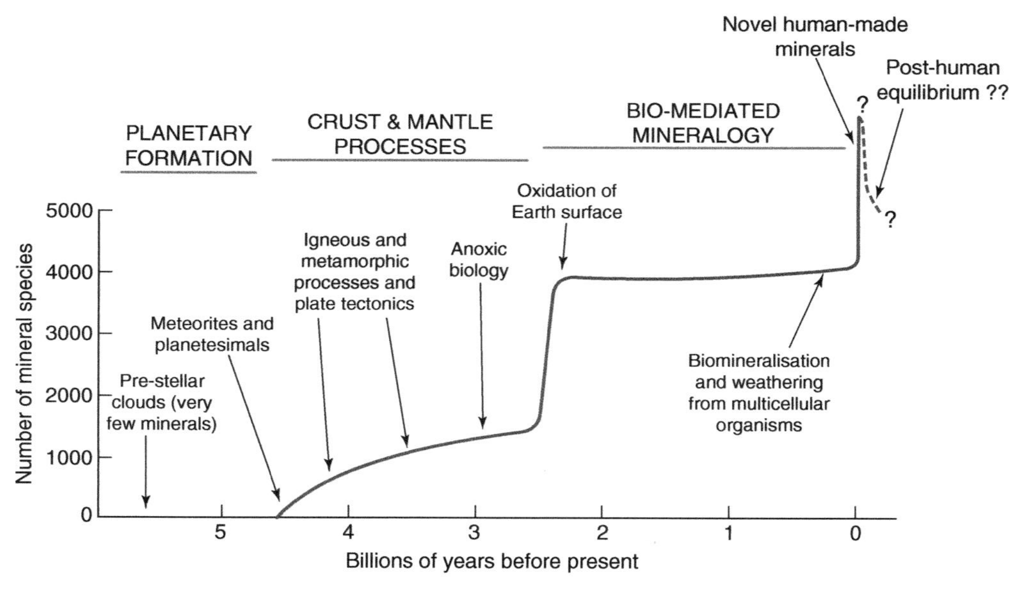

34. The growth of mineral species from before the birth of our solar system to the present day.

what one might regard as reality, rather than formality, as the International Mineralogical Association (the body that oversees the identification and validation of new minerals) specifically excludes human-made minerals from its official list.

Nevertheless, humans have certainly introduced mineralogical novelties to this planet. One might begin with metals. There are only a few more or less uncombined, or 'native', metals in nature. Gold, silver, and the platinum group elements may occur naturally as metal, and occasionally copper, while there is a little native iron too, mainly found as meteorites. Otherwise, naturally occurring metals are almost always bound up as chemical compounds. Humans began to separate out metals such as copper, lead, tin, and iron, and to combine them in new alloys, such as bronze and brass, several thousand years ago. In more recent times, especially since the Industrial Revolution, many more metals (ultimately, all of them) were made in separated form in the laboratory, and some began to be produced in industrial quantities in the burgeoning factories.

Aluminium is an example. It is almost unknown as an uncombined metal in nature, being only rarely found, and then in tiny amounts, in very reducing environments (such as in submarine cold seeps). It was produced artificially in 1829, but remained a rarity until the mid-20th century, when its production began to grow sharply, up to its current production levels of about 35 million tons per year, the cumulative amount produced now totalling something of the order of 500 million tons—enough to cover the USA and part of Canada in standard thickness kitchen foil (Figure 35). Other metals, absent or virtually absent in nature, such as titanium, are now also made in large amounts, while the production of iron is now about a billion tons annually.

A wide variety of mineral compounds rare or unknown in nature have been devised, too, by the materials chemists, for a variety of purposes. These include tungsten carbide, which forms the ball in

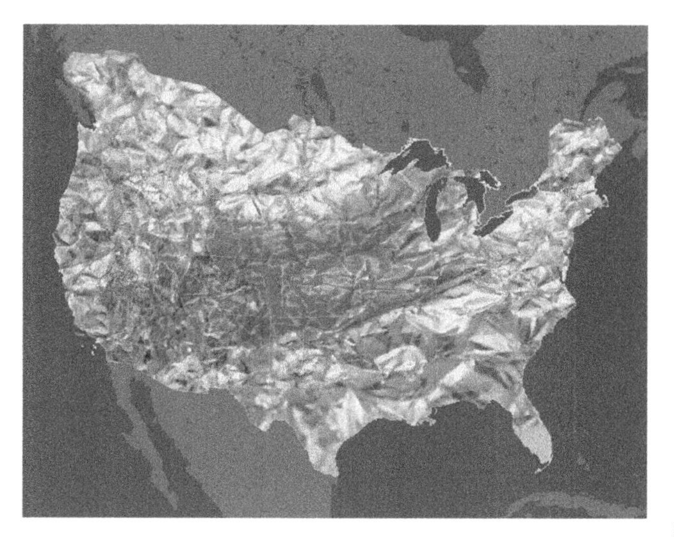

35. North America covered with the total global human-made production of aluminium, expressed as standard kitchen foil.

many ballpoint pens; a variety of synthetic rubies, used in lasers; and boron nitride, harder than diamond and manufactured in hundreds of tons per year to use as an abrasive. There are also *mineraloids*—compounds of chemical composition that can vary within limits, such as plastics and glass. The use of plastics has grown since the mid-20th century to now be some 300 million tons per year (i.e. an amount equalling the mass of the entire human population). Much of these new minerals and mineraloids are not recycled and either end up in landfills or are dispersed far into the environment, including to distant beaches and the bottom of the oceans.

Because they are new, the geological longevity of these human-made minerals cannot be precisely judged. However, some idea of how well they might fossilize can be gained by considering how they may behave chemically in the underground environment of buried strata. Aluminium metal resists corrosion in surface environments not because it is unreactive (quite the converse) but because it almost instantly develops a thin oxide layer that resists further chemical weathering. Buried underground, in reducing conditions, this oxide layer might be expected to break down, and the aluminium would corrode. By that time, though, any buried aluminium object (a can, perhaps) might be expected to have left a permanent impression on the stratal surfaces it is encased in. And, as the aluminium metal itself slowly dissolves, it might be expected to react with the neighbouring minerals to give products such as new clay minerals, that might line the impression of the can in the rock, and in this fashion perhaps survive for geological ages (until, perhaps, it may be excavated and puzzled over by some geologist in the far future).

Human-made minerals are produced in laboratories and factories around the world, with many new forms appearing every year. How many of these new forms are there? Materials sciences databases now being compiled suggest that more than 50,000 solid, inorganic, crystalline species have been created in the laboratory. Almost certainly, they represent the greatest increase in mineral diversity on Earth since the Great Oxygenation Event of the early Proterozoic eon, almost two and a half billion years ago.

The mineral transformation of Earth is one of a number of interlinked changes—to rocks and fossils, as we shall see, and to biodiversity and climate and much else—that has been described in terms of the 'Anthropocene Epoch'. This is a still informal term that reflects the idea that humans are now the main drivers of the Earth's geology, and have already—no matter what else will happen—changed the course of this planet's history. The term was coined by one of the world's great scientists, Paul Crutzen, at a conference in Mexico in 2000. His fellow scientists were talking of the environmental changes taking place during these recent years of the Holocene (the epoch of post-glacial time in which we still, formally, live). He became increasingly frustrated with this discussion, until he could contain himself no longer. 'We are no longer living in the Holocene', he burst out. 'We are living in…' (he paused, seeking to find a word that expressed what he felt) 'the Anthropocene'. Study of this concept is currently growing apace.

However, Paul Crutzen is not a geologist, but an atmospheric chemist. As a child he survived harsh times in wartime Holland, including the terrible *Hongerwinter* ('hunger winter') of 1944, when some of his schoolfriends died of starvation. He went on to work on some of the largest possible themes of Earth and atmospheric science. He was centrally involved in developing the theory of the nuclear winter, something that has fortunately not (yet) happened; and of the mechanism of human-driven ozone destruction, for which he won a Nobel Prize in 1995. Yet his championing of the Anthropocene concept may prove his most enduring legacy—one that will be written, indeed, in rock.

Human-made rocks

The new minerals are becoming distinctive components of sedimentary strata now forming. Some are also becoming components of new rock types specifically made by humans. Bricks are among the earliest, and still among the most volumetrically significant of these. Thousands of years ago, these were of shaped, sun-dried mud, and so they differed little from the original sediment. However, as techniques such as firing in kilns came into play, what is now produced is essentially a flash-metamorphosed rock: temperatures typically reach levels sufficient to begin to melt the sediment. In these conditions, minerals that are a little uncommon in nature, such as mullite, an aluminium silicate, crystallize to become a common component of the rock. Something like a trillion bricks are now made a year.

Similar new rocks, of more or less novel and distinctive mineralogy, are produced as artificial materials. They include ceramics, cement, and concrete. The manufacture of concrete involves the large-scale crystallization of minerals such as as ettringite and portlandite, which in nature are somewhat obscure components of a few metamorphosed limestones. Concrete itself as a novel rock type is distinctive geologically—especially when components such as fly ash (tiny mineral or carbon spheres produced by the industrial burning of hydrocarbons) are incorporated within it. Although made since Roman times in some form or other, it is only since the mid-20th century that its use has skyrocketed towards its current position of being the building material of choice throughout the world (Figure 36). In all, something like half a trillion tons has now been produced, which is in the order of 1 kg for every square metre of the Earth's surface, land, and sea.

Rock production by humans is now planetary in scale. It is concentrated in urban areas, but there are significant amounts within transport networks, dams, flood defences, and other structures. And it is in the complex structures, at all scales, that the true novelty of human-made rocks is most clearly expressed. These vary from accumulations that may be regarded as human-made strata, to be delineated on geological maps, to more sophisticated constructions that we might call 'technofossils'.

The new rock formations

When geologists analyse rock strata, a key process is their formal classification. Here rocks are classified into units large enough to be shown on geological maps. Key to this is the setting up of rock formations, units with defined boundaries that can be recognized and traced across country.

Human-made strata are now on a scale sufficient to be also shown on geological maps, as forms of what may be termed 'artificial ground'. Such features are of various geometrical and textural forms—and can build up substantial, positive, topographic features. Large amounts of soil and rock, for instance, are moved to build up structures such as dams and embankments, or simply in the form of waste heaps from mines and quarries of various

36. An accumulation of the geologically novel rock: concrete. (San Francisco, USA.)

kinds. There are also negative topographic features—the mines and quarries themselves—which may be equally substantial. The Hibbing iron ore quarry in Minnesota, North America, for example, is 8 km long, 3 km wide, and reaches 180 metres in depth—almost a billion tons of rock have been extracted from it (the equivalent, it is said, of boring a tunnel through to the centre of the Earth and then out to the other side).

And, once the minerals have been extracted from such quarries, the resulting depressions are often filled with other materials, notably the rubbish that we throw away every day, as we replace one type of stratum for another.

The total amount of such material being moved and redeposited by humans—which includes the soils stripped away during deforestation and farming, and the sediment building up behind large dams—is now estimated to be several times greater than is moved by the natural forces of erosion and deposition. In recent decades, such human-driven, mass sediment movement has also moved out to sea, for instance by the bottom trawling (submarine ploughing, in essence) that now changes the texture and fauna of the sea floors of most continental shelves across the world (and is moving down the continental slopes). This kind of topographic reshaping is creating distinctive new strata, which may be fossilized far into the future, as a distinctive new addition to the rock cycle. And the reshaping extends very deep.

Subterranean rock transformations

In the Cambrian Period, half a billion years ago, animals grew muscles and learnt to burrow down a few centimetres into the sediment of the sea floor, beginning in this way to transform the world. Today, champion burrowers in the ocean include the shrimp *Callianassa*, whose burrow systems may extend down a couple of metres. On land, wolf and fox burrows may go down twice as far, while the Nile crocodile, when preparing to aestivate, may dig down as much as 12 metres through the river muds.

Humans, though, have outdone these feats by orders of magnitude. Even in the Bronze Age, people armed only with antlers and boulders could dig a couple of tens of metres through solid rock to seek (and find) copper ore, as at Parys Mountain in Anglesey. In more recent times, humans commonly dig down several hundred metres into solid rock, to extract entire coal seams or mineral veins. Sometimes they go even deeper, as in the almost 4-km-deep gold mines of South Africa (with plans to extend the workings to 5 km below ground).

Even if humans themselves go no further, by drilling through the rock, they do commonly reach considerably deeper, to several kilometres below ground. There are very many of these boreholes. The number of oil boreholes alone (not counting those for minerals or, now, for fracking for gas) in North America alone has been estimated at 5 million km in total, while the figure for the world may be something like 50 million km—equivalent to the length of the world's road network (or the distance that separates Earth from Mars).

This artificial network of underground rock punctures is quite unique in the four and a half billion years of Earth history, especially in being produced by a biological organism. The punctures themselves are substantial things, not to speak of the translocation of materials (e.g. oil, upwards; and drilling mud, downwards) that they have expedited. The relatively new process of fracking to extract gas from carbon-rich mudrock extends this process to fracturing large areas of mudrock hundreds or thousands of metres below the ground, while injecting large amounts of sand grains to hold the fractures open while the gas can seep out to be collected. This is in effect producing a new, distinctive rock type.

Unlike human-made surface structures, this three-dimensional network of boreholes and mines, being so deep, is effectively beyond the reach of erosion, and so it is as near eternal as anything on Earth can be. Parts, for sure, will eventually (after tens or hundreds of millions of years) rise to the surface to be eroded away, and parts may be effaced by metamorphism or underground magmatism, but human subterranean explorations will form one of the more enduring parts of our rock-bound legacy.

The most spectacular examples of these human-made underground rock changes do indeed involve metamorphism—caused by shock and magmatism. Beneath such places as the Yucca Flats in Arizona in North America as well as Semipalatinsk and Degelen Mountain in Kazakhstan, over 1,350 boreholes were drilled to carry nuclear bombs 1 km or so below ground. When they were detonated, they often left a circular crater, hundreds of metres across, at the surface. The real legacy, though, lies far below ground, amid the masses of shock-brecciated and radioactive rock, with substantial amounts of melt pooling following larger explosions.

These particular relics give much pause for thought, not least as geological souvenirs of humankind. However, our constructional legacy may—perhaps thankfully—be yet more extraordinary.

Technofossils

When palaeontologists examine fossils within rock strata, a basic division is between *body fossils*, such as bones and teeth, and shells; and *trace fossils*, such as tracks, pawprints, and burrows. Humans can leave tracks and footprints too, on a beach or on a muddy field; and, as we have seen, they can burrow deeply too. But there are other kinds of trace fossils: the ancient volcanic soil layers of the Canary Islands, for instance, are often crowded with fossilized acorn-shaped nests, painstakingly constructed by burrowing wasps from carefully size-selected pieces of pumice. One can even occasionally find fossilized termite nests too, that are up to 3 metres tall and range in age back to the Jurassic, over 150 million years ago.

The human equivalents of these are the roads and buildings of our towns and cities. They are also made out of geological materials—of reconstituted rock, as it were—and are eminently fossilizable (if they are in the right location, as we shall see later). But, they are on a huge scale, and they are almost infinitely more intricate than any termite nest (which are not, one must recall, unsophisticated). And, they are evolving very quickly.

The scale is visible when we see a satellite image of any populous and industrialized country, for the megacities of the world now each cover hundreds of square kilometres. Their structures reach tens and sometimes hundreds of metres into the sky, while their foundations, including water and waste pipes, electricity cables, metro systems, and skyscraper pilings, can reach downwards into soil and rock by an equivalent amount.

The complexity of megacities not only relates to their roads and buildings, remarkable geometrical constructions as they are. They are filled with objects that we make for our needs and our pleasures: chairs, tables, cutlery, carpets, books, mirrors, cupboards, pictures, televisions, computers, and an almost infinite number of other objects, while outside in the driveway there will be one or more cars (not to mention the aircraft parked on the runway outside the city). All these objects are technological items shaped by us from materials we have grown or, more commonly, extracted from rock in the ground. They are all, or nearly all, potentially fossilizeable. Hence, one might call them—with an eye to the future—'technofossils' (Figure 37). They will likely be a truly extraordinary part of our rockbound legacy.

One remarkable property of such new human-made constructions is the intricacy of the paths taken by their components. In normal sedimentary or fossil systems, the movement of material is either by gravity or by waves or currents (in assembling sedimentary rock strata) or by biological migrations (in assembling fossil assemblages). The components of our new, enormous, trace fossil assemblages were brought together through different, and often much longer and more intricate, travel paths, typically powered by

37. A technofossil in the making—the next day, it was 'fossilized' by being covered with a layer of road tarmac.

the burning of fossil fuels. A single mobile phone, for instance, contains dozens of elements, including gallium, indium, niobium, palladium, gold, and silver, sourced from all over the world and assembled in one small structure, just a few centimetres across.

Humans have long made tools—from, say, pieces of bone or flint—even before our own species, *Homo sapiens*, came into existence. Even some non-human animals—chimpanzees, crows—can use tools in a simple fashion. And human tools have evolved through time, as archaeological and historical studies have clearly showed. But over the past few decades there has been an explosive radiation of technology, and the tools that it has produced, which is without any precedent on this planet. Indeed, technology is now so central to our lives (without it, only a small fraction of our current numbers could exist) that it has been placed as a central component of what the geologist Peter Haff has termed the 'technosphere'—an offshoot of the biosphere.

The evolution and diversification of technology and its material products is now proceeding apace and accelerating, entirely decoupled from the biological evolution of the species (ours) that has created it. It enables by far the greater part of the alteration that we are currently making to the rock structure of the Earth. And it is quite uncertain where this revolution is taking us. Some have even proposed that human intelligence may be outstripped by silicon intelligence, with quite unknown implications for the geological evolution of this planet. We are, quite certainly, living in interesting times. And rock will, very likely, never be the same again.

Further reading

Fortey, Richard. 2005. *Earth: An Intimate History*. Harper. Excellent on the various phenomena—volcanoes, earthquakes—that are by-products of the working of Earth's plate tectonics engine. Richard Fortey's *Hidden Landscape* (Bodley Head, 2010) also gets nicely under the skin of ancient landscapes.

Hazen, Robert. 2012. *The Story of Earth*. Penguin. A fine account of our planet seen through the eyes of this distinguished and imaginative mineralogist.

Levi, Primo. 1984. *The Periodic Table*. Abacus Books. Among the most beautiful popular science books ever written, its chapters on gold and lead and how they relate to the rocks of the Earth are reason enough for its inclusion here.

Mantell, Gideon. 1836. *Thoughts on a Pebble*. This remains a charming and still informative exploration of a single flint pebble out of the English Chalk; though out of print for a century and more, it now easy to access on the internet. I essayed a kind of reprise, within the *quite* different world of Welsh slate, in *The Planet in a Pebble* (Oxford University Press, 2010).

Nield, Ted. 2007. *Supercontinent: Ten Billion Years in the Life of Our Planet*. Granta Books. Highly readable and nicely quirky account of the working of past (and future) plate tectonics. His *Underlands* (Granta, 2014) eloquently explores the link between rock and people.

Redfern, Martin. *The Earth: A Very Short Introduction.* Oxford University Press. Excellent and remarkably comprehensive on the inner workings of the Earth.

Welland, Michael. 2008. *Sand.* Oxford University Press. The never-ending story of this inexhaustively various material, marvellously told.

Index

A

B

C

D

E

F

M

N

O

P

Q

R

S

T

U

V

W

X

Y

Z

DESERTS

A Very Short Introduction

Nick Middleton

Deserts make up a third of the planet's land surface, but if you picture a desert, what comes to mind? A wasteland? A drought? A place devoid of all life forms? Deserts are remarkable places. Typified by drought and extremes of temperature, they can be harsh and hostile; but many deserts are also spectacularly beautiful, and on occasion teem with life. Nick Middleton explores how each desert is unique: through fantastic life forms, extraordinary scenery, and ingenious human adaptations. He demonstrates a desert's immense natural beauty, its rich biodiversity, and uncovers a long history of successful human occupation. This *Very Short Introduction* tells you everything you ever wanted to know about these extraordinary places and captures their importance in the working of our planet.

www.oup.com/vsi

ONLINE CATALOGUE

A Very Short Introduction

Our online catalogue is designed to make it easy to find your ideal Very Short Introduction. View the entire collection by subject area, watch author videos, read sample chapters, and download reading guides.